REVISE AQA GCSE (9–1) Geography

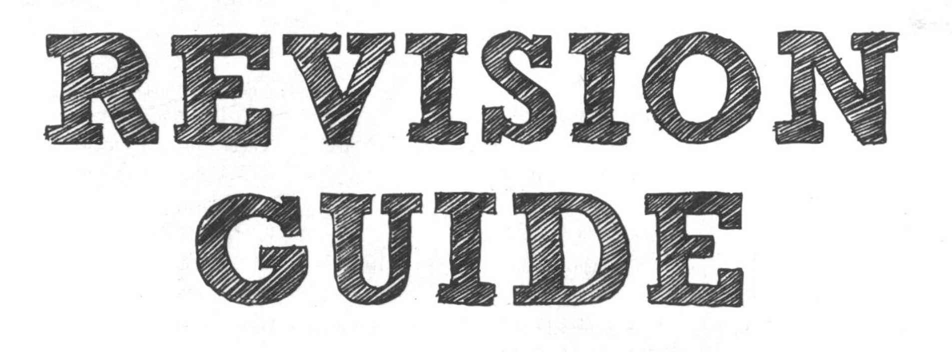

Series Consultant: Harry Smith

Author: Rob Bircher

Contents

1-to-1 page match with the AQA Geography Revision Workbook ISBN 9781292131313

AQA publishes Sample Assessment Material and the Specification on its website. This is the official content and this book should be used in conjunction with it. The questions in Now try this have been written to help you practise every topic in the book. Remember: the real exam questions may not look like this.

Natural hazards

A natural hazard is a natural event that is a threat to people and to property. The risk of this threat increases when a natural hazard occurs where lots of people live.

Types of natural hazard

Earthquakes and volcanic eruptions are the most significant **tectonic hazards** in terms of risk to people's lives and property.

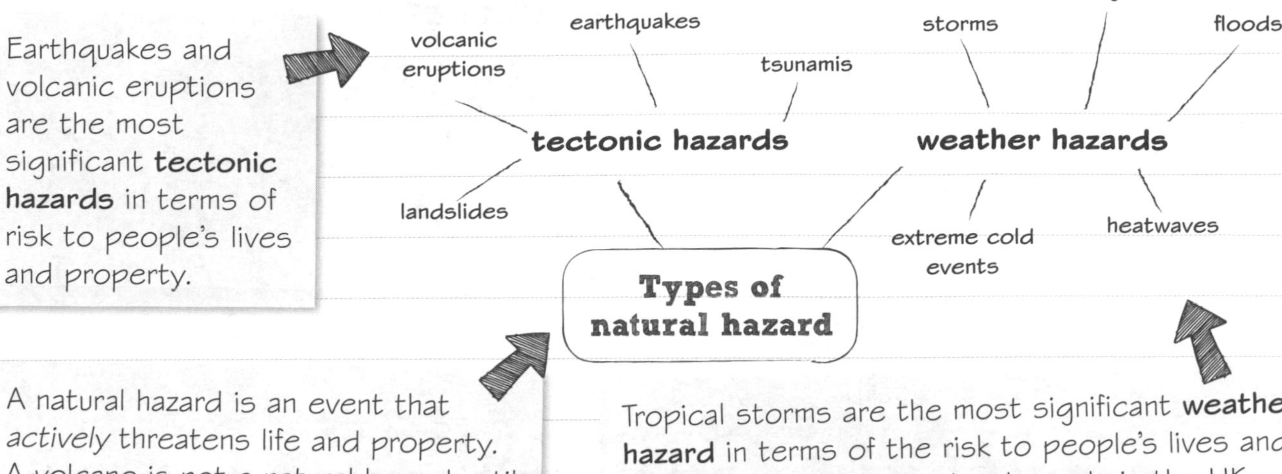

A natural hazard is an event that *actively* threatens life and property. A volcano is not a natural hazard until it erupts or threatens to erupt.

Tropical storms are the most significant **weather hazard** in terms of the risk to people's lives and property. We have weather hazards in the UK, including droughts, heatwaves and storm events.

Hazard risk

Different factors affect how likely it is that people will be hit by a natural hazard.

Type of hazard	Some hazards are difficult to predict – this makes them more risky
Economic development	Richer countries can afford monitoring, prediction, protection and planning to reduce risk
Population growth	When a country has a rapidly growing population, people may have to live in hazardous areas, such as mountain slopes
Severity	Some natural hazards are more severe than others of the same type – for example, a higher magnitude, a greater intensity

This answer has simply given two factors. This is the right approach as this is all that the question asked for.

Worked example

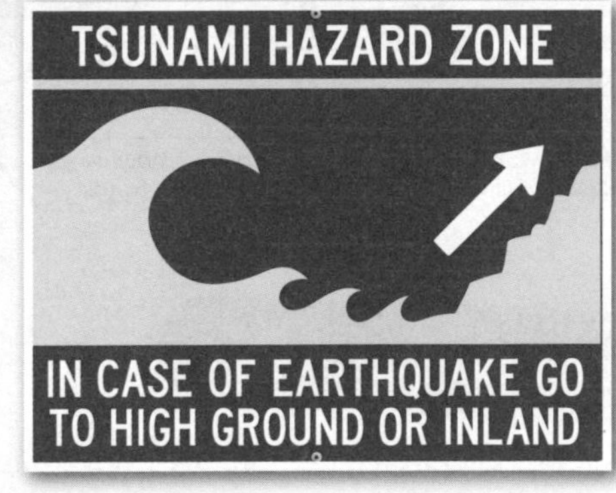

Study the image above, which shows a warning sign in an area at risk of tsunami hazards.

Give **two** factors that could increase tsunami hazard risk in a coastal area. **(2 marks)**

If the coast was close to a destructive plate margin. If the coastline was heavily populated.

Now try this

Which **one** of the following is a natural hazard?

☐ **A** An earthquake in an uninhabited region
☐ **B** Air pollution in an industrial city
☐ **C** A storm surge along a heavily populated coastline
☐ **D** Loss of coral reef biodiversity due to a rise in seawater temperature **(1 mark)**

Plate tectonics theory

Mapping the distribution of volcanic eruptions and earthquakes provided evidence for the theory that the Earth's crust is divided into slowly moving tectonic plates.

Global distribution of earthquakes

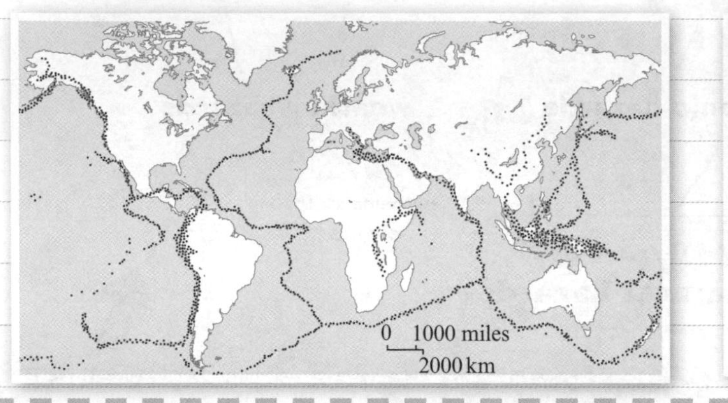

Global distribution of volcanoes

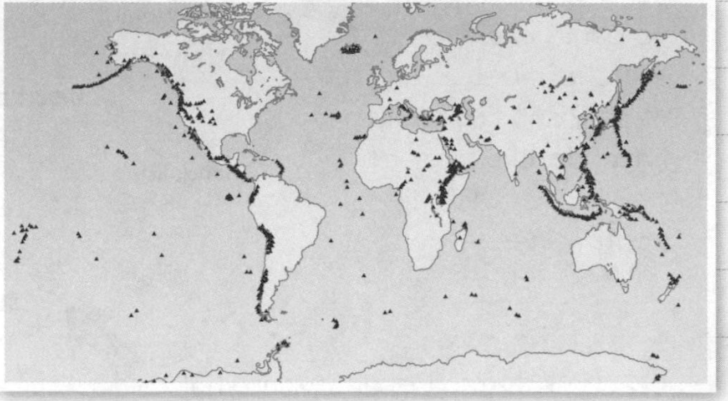

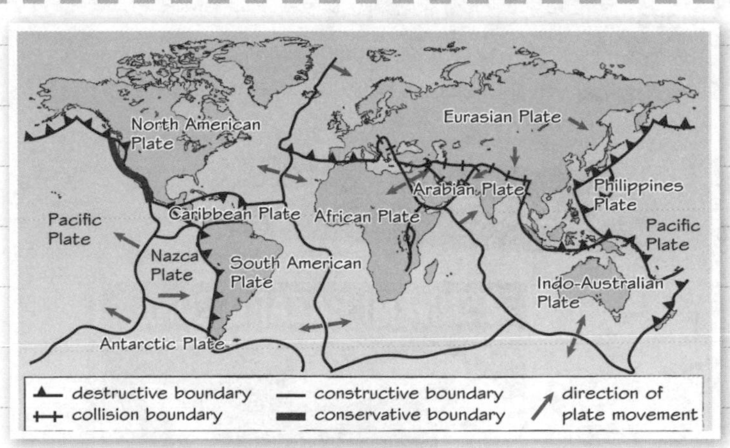

Earth's main tectonic plates

The Earth's crust, the **lithosphere**, is broken into plates. There are two types of crust:

- **continental crust** – thicker and less dense
- **oceanic crust** – thinner and more dense.

The reason plates move is not completely understood, but one theory is that they are driven by **convection currents** in the mantle layer underneath the Earth's crust.

Worked example

Outline **one** way in which plate tectonics theory explains the global distribution of earthquakes and volcanoes. **(2 marks)**

Earthquakes and volcanoes are mainly distributed along plate margins – where two (or more) plates meet. Plate tectonic theory says that plates move against each other and this generates vast amounts of energy. This energy is what causes volcanoes and earthquakes.

This is a good answer because it makes a point and then develops it.

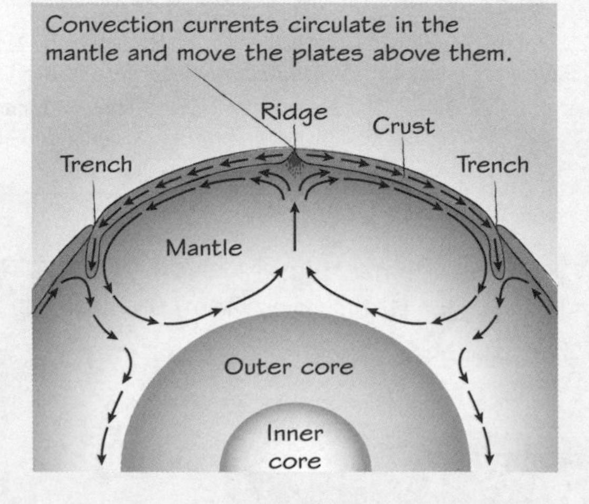

Now try this

1 Describe the global distribution of earthquakes shown on this page. **(2 marks)**

2 A few volcanoes occur far away from plate margins at 'hotspots'. Explain how these volcanoes provide support for plate tectonics theory. **(4 marks)**

Plate margin processes

Physical processes that take place at different types of plate margin (**constructive**, **destructive** and **conservative**) lead to earthquakes and volcanic eruptions.

Constructive plate margin

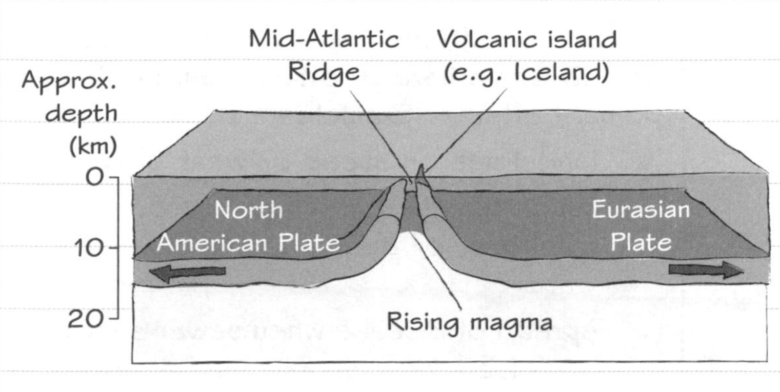

A constructive margin is when plates move apart and magma rises to fill the gap.

- Constructive margins are usually found under the sea.
- Ridges and **shield volcanoes** are formed by the build-up of **magma**. Fluid spreads out a long way forming gently-sloping shield shape.
- Shallow, minor earthquakes can occur along cracks in the plates as they move apart.

Destructive plate margin

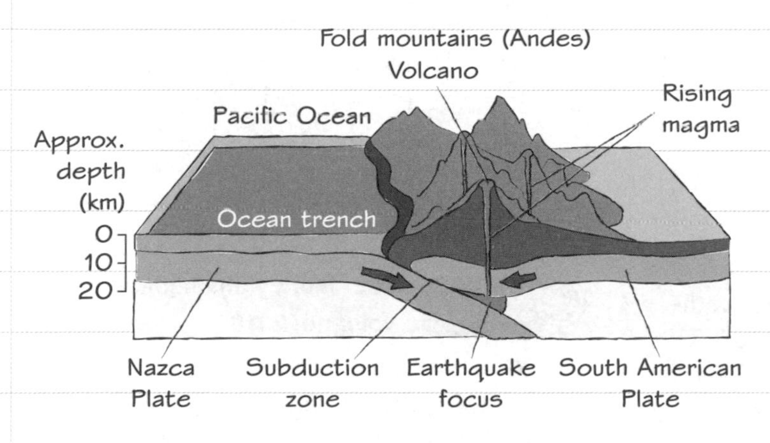

A destructive margin is when plates move together. The lighter continental crust stays on top while the denser oceanic crust is pushed (subducted) into the mantle where it melts.

- Rising magma from the melting crust can cause **composite volcanoes**.
- Energy builds up in the subduction zone and is released as violent eruptions which often feature pyroclastic flows. These form steep-sided volcanoes made up of layers of ash and lava.

Conservative plate margin

San Andreas Fault
North American Plate
Faster movement
Pacific Plate

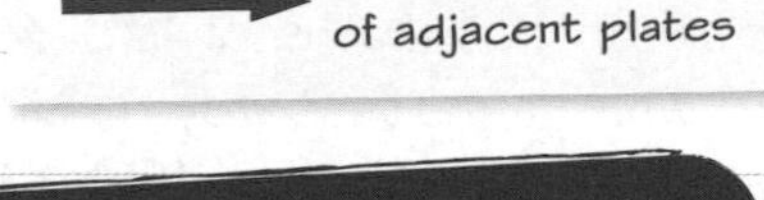

A conservative plate margin occurs when plates slide past each other.

- No crust is being created or destroyed, therefore **no volcanoes** are formed.
- Friction between the two plates can make them 'stick'; pressure then builds until the plates 'unstick', causing violent earthquakes.

Worked example

Give **two** ways in which earthquakes are triggered by plate movements. **(2 marks)**

At destructive plate margins, energy builds up in the subduction zone and is released as earthquakes. At conservative plate margins, if the plates get stuck, pressure builds up and is released as earthquakes when the plates move.

A good answer with two ways described in full.

Now try this

Give **two** ways in which volcanoes are caused by plate movements. **(2 marks)**

Tectonic hazards: effects

You need to know **named examples** of the primary and secondary effects of a tectonic hazard that come from **two areas of contrasting levels of wealth.**

Primary effects

Primary effects happen immediately as a result of a tectonic hazard. Examples include:

- ✓ deaths and injuries
- ✓ destruction of/damage to buildings
- ✓ destruction of infrastructure, such as phone/electricity cables, roads, ports, railways.

How serious primary effects are depends on a mix of physical and human factors, such as an earthquake's magnitude (measured by the Richter scale) or whether the hazard occurs in a densely or sparsely populated area.

Secondary effects

Secondary effects are the after-effects of a tectonic hazard, often resulting from primary effects. Examples include:

- ✓ landslides on steep or weak slopes
- ✓ tsunamis when earthquakes occur offshore, or due to landslides into water
- ✓ spread of disease when sewer systems and clean water supplies break down
- ✓ fires caused by fractured gas pipes.

Secondary effects may have more impact in Lower Income Countries (LICs) than in High Income Countries (HICs), as LICs may lack funds to monitor and predict hazards, or to prepare people or protect infrastructure.

Worked example

Named example

Using named examples, describe how the primary effects of a tectonic hazard have varied in two areas of contrasting levels of wealth. **(6 marks)**

On 22 February 2011, a 6.3 magnitude earthquake hit the city of Christchurch, New Zealand. New Zealand is a wealthy country with a Gross National Income (GNI) per capita of US$40020. The primary effects included the deaths of 185 people, most of whom died when a multi-storey building collapsed; 7100 people were injured; 100000 buildings were damaged. The Haiti earthquake happened on 12 January 2010, with a magnitude of 7.0. Haiti is a very poor country, with a per capita GNI of just US$810. As in New Zealand, the epicentre was close to a major city, Port-au-Prince. In total, around 230000 people were killed and 180000 buildings were destroyed. As in Christchurch, roads were severely damaged and the main port was also destroyed.

Fact file: Eyjafjallajökull volcano, 2010

The Iceland volcano Eyjafjallajökull erupted from March to May 2010. During April, the eruptions began to release enormous amounts of ash into the atmosphere. It was advised that flying through the ash could damage aircraft engines. As a result:

- 104 000 flights were cancelled, affecting 10 million passengers. Losses to the aircraft industry have been estimated at £1.1 billion.
- Imports of fresh flowers from Africa to Europe were cancelled. Kenya lost around US$24 million from cancelled flower exports.
- Car production in Japan came to a standstill as parts made in the EU could not be shipped in time.

Try this question for your own tectonic hazard named examples – maybe you have studied volcanoes instead of earthquakes, or different earthquakes?

Now try this

Which **one** of the following describes the impacts of Eyjafjallajökull listed in the Fact file?

☐ **A** Primary effects – caused by the volcano

☐ **B** Secondary effects – an indirect result of the volcano

☐ **C** Immediate responses – responses to the eruption itself

☐ **D** Long-term responses – went on for a long time after the eruption

(1 mark)

Tectonic hazards: responses

You need to know **named examples** of immediate and long-term responses to a tectonic hazard in **two areas of contrasting levels of wealth**.

Responses

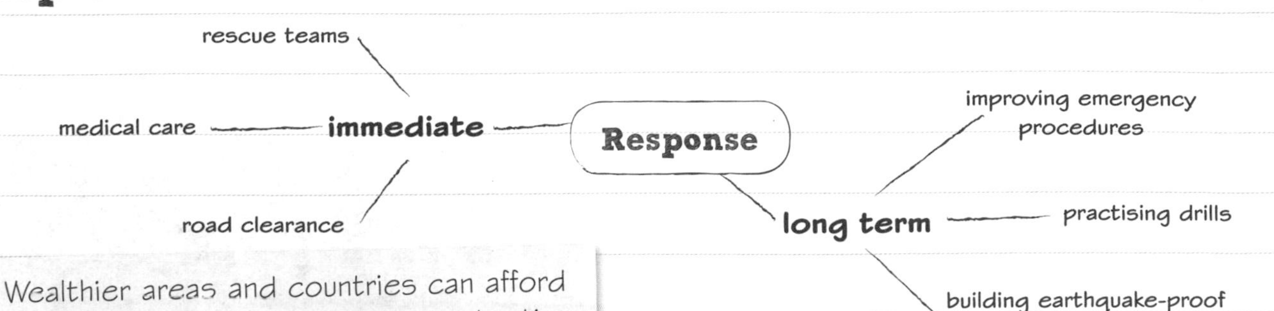

Wealthier areas and countries can afford better monitoring, prediction, protection and planning than less wealthy areas.

Worked example

Named example

Using named examples, discuss how the long-term responses to a tectonic hazard have varied between two areas of contrasting levels of wealth. **(6 marks)**

The earthquake that struck Christchurch, New Zealand, on 22 February 2011 caused $40 billion of damage. Eighty per cent of the water and power infrastructure was destroyed and 10 000 buildings had to be demolished. The long-term responses were thorough. By August, 80 per cent of roads had been cleared of debris, and water supplies and sewers were fixed for all residents. Areas were zoned so it was clear which land was safe to rebuild on and which buildings were secure. Most Christchurch residents, who had moved out after the quake, could then return to their homes. The long-term responses to the Haiti earthquake of 12 January 2010 show some contrasts to those of Christchurch. There were few earthquake-resistant buildings in this very poor country, so huge numbers of houses were destroyed. Despite long-term aid from charities, a year after the quake 98 per cent of roads were still not usable because of rubble, 1 million people were still living in camps because houses had not been rebuilt and 70 per cent of people had not been able to return to their work and were depending on international aid.

Named example

Fact file: Japan earthquake and tsunami, 2011

- When: 11 March 2011
- Earthquake magnitude: 9.1
- Maximum tsunami height: 39 metres
- 20 000 dead, 5000 injured
- 125 000 buildings destroyed
- 6 million houses were left without electricity, 1.5 million were without water
- Nuclear reactors at Fukushima Nuclear Power Plant were damaged; a 12-mile exclusion zone was set up around the plant
- 100 000 volunteers were mobilised to help with disaster relief
- The Japanese army cleared roads of rubble in two days, allowing relief to get to cut-off communities
- 450 000 people were moved to emergency shelters, although lack of electricity meant they were cold
- By November 2011, all electricity, water supply and telephone line repairs were completed
- Special development zones were set up to attract investment for redevelopment
- The government agreed a fund of US$200 billion for rebuilding over the next 10 years

This is a good answer because it focuses on long-term responses only.

Now try this

From the Fact file above, identify **one** immediate response to the 2011 Japanese earthquake and **one** long-term response. **(2 marks)**

Had a look ☐ Nearly there ☐ Nailed it! ☐

Living with tectonic hazards

Why do people continue to live in areas that are at risk from a tectonic hazard?

Worked example

Study the photo and read the text below. Eight million people live in the San Francisco Bay area today.

Suggest **two** reasons why people continue to live in San Francisco, despite it being at risk from earthquakes. **(2 marks)**

The earthquakes in San Francisco happened a long time apart – 83 years – so people do not think an earthquake is likely to affect them. Plate margins are often found along coastal areas. San Francisco is a port city. Being on the coast is a major reason why it has been economically successful. People live there for well-paid jobs despite the tectonic hazard risk.

San Francisco, 1906

San Francisco, California, sits on the San Andreas Fault and is prone to earthquakes.

- In 1906, an earthquake killed 3000 people and destroyed 28 000 buildings. In total, 225 000 people were made homeless.
- In 1989, a 6.9 magnitude earthquake killed 67 people and caused US$5 billion of damage. Following the earthquake, all older buildings were reinforced.

Living in hazardous areas

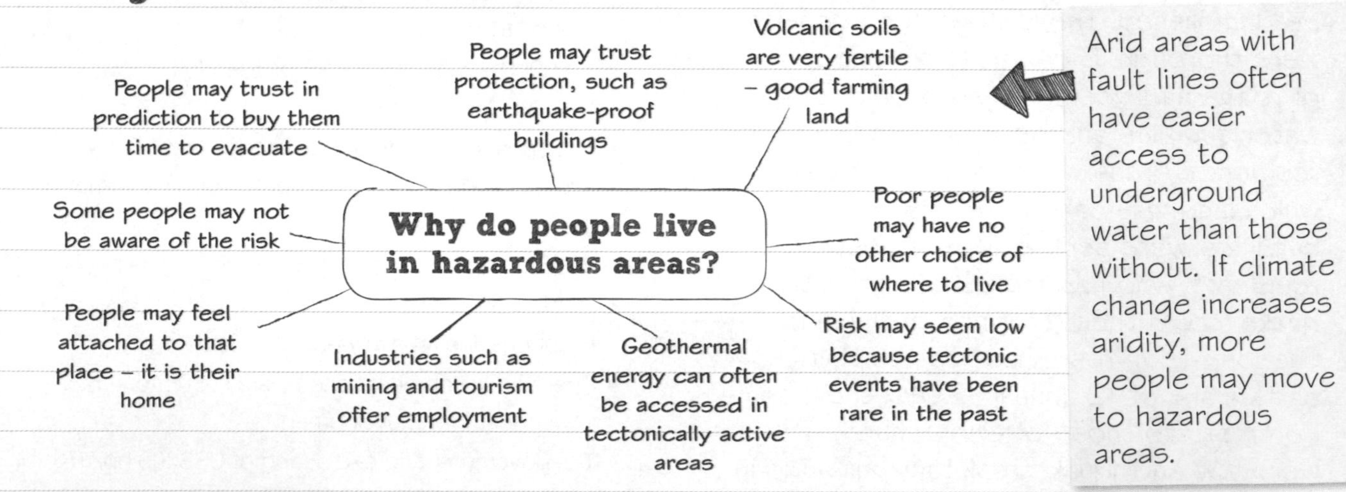

Arid areas with fault lines often have easier access to underground water than those without. If climate change increases aridity, more people may move to hazardous areas.

Now try this

This satellite image shows Mount Vesuvius. About 600 000 people live around this active volcano in a 'red zone', which is at very high risk in the event of an eruption.

Suggest **two** reasons why so many people continue to live in this high-risk area. **(2 marks)**

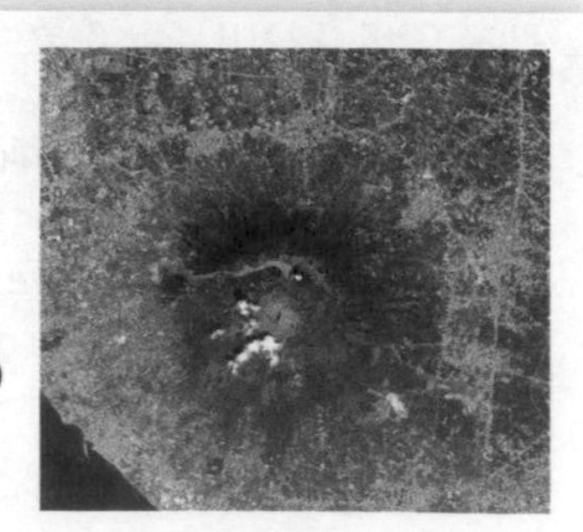

Tectonic hazards: reducing risks

Monitoring, prediction, protection and planning can help reduce the risks from a tectonic hazard.

Monitoring and prediction

Several instruments can monitor seismic activity to help predict when tectonic hazards may occur.

1. Tiltmeters and lasers measure change in the slope of the land, showing deformation of the surface.
2. Seismometers measure the seismic waves produced by earthquakes. A series of smaller quakes may come before a big one.
3. GPS and satellite images monitor the bulging shape and changes in ground heat that indicate a volcanic eruption is coming.
4. Chemical sensors measure changes in gases (carbon dioxide and sulphur dioxide) that indicate magma activity below ground. Releases of radon gas from cracks in the Earth's surface may indicate an earthquake is likely.

Protection

Buildings can be reinforced against earthquakes.

- Installing a ring beam (band of concrete) at roof level to stop walls falling outwards
- Very strong framework in skyscrapers
- Strengthening walls
- Making foundations from rubber and steel which can move slightly
- Digging deeper foundations
- Reinforce gas and water pipes so they do not break

It is hard to protect buildings from volcanic eruptions, although sometimes channels are dug to divert flows of lava. It is also hard to monitor and predict earthquakes – there are no really reliable indicators.

Worked example

The image opposite is of a **Global Positioning System (GPS)** station located on a plate boundary in the USA.

Explain why prediction of tectonic hazards is better in HICs than in some LICs and Newly Emerging Economies (NEEs). **(4 marks)**

GPS prediction technology is expensive to install in the first place, but what LICs often have most problems with is the cost of maintaining the equipment over time.

Governments run volcano prediction projects, but in LICs governments cannot afford to pay high wages. The engineers and technical analysts needed for monitoring and analysis may go to work in better-paid jobs in the private sector.

This answer makes two supported points, which help with the explanation.

Planning

- Hazard mapping identifies areas most at risk from tectonic events.
- Evacuation plans set up routes for people to escape to places where they can get shelter.
- Citizens and emergency services practise drills: what to do in the event of a tectonic hazard.

Now try this

Suggest **one** way in which planning can help reduce the effects of an earthquake.

(1 mark)

Global atmospheric circulation

The general atmospheric circulation model helps explain how the **global atmospheric circulation** influences patterns of weather and climate – including weather hazards.

Circulation cells

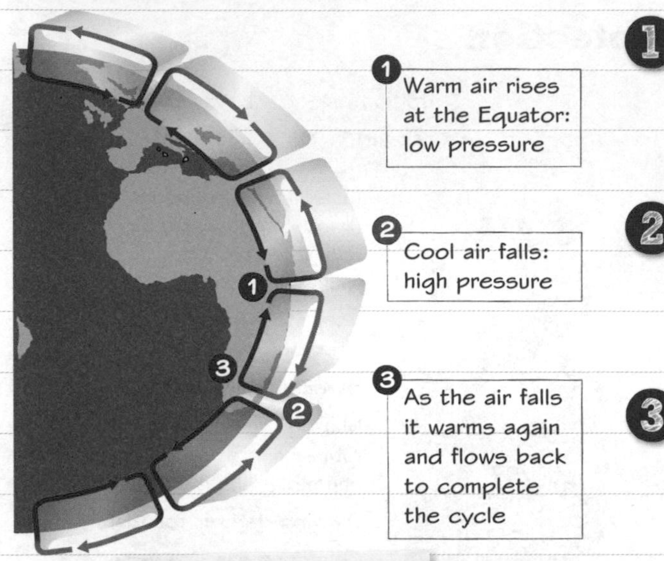

Air flow in circulation cell

1. The Earth receives its energy from the Sun. The Sun's UV rays create intense heat energy – most intense at the Equator – powering a huge circular air movement called a **circulation cell**.
2. Hot air at the Equator rises 15 km into the atmosphere. This air cools and travels north and south to around 30° of latitude, where it sinks. Where the cells meet, energy gets transferred.
3. When warm air rises, it creates low pressure. There is a **belt of low pressure** around the Equator. When cool air falls at 30° of latitude north and south, it creates a **belt of high pressure.** High-pressure conditions have clear skies with little precipitation; these are arid areas.

Affect of circulation cells on surface winds

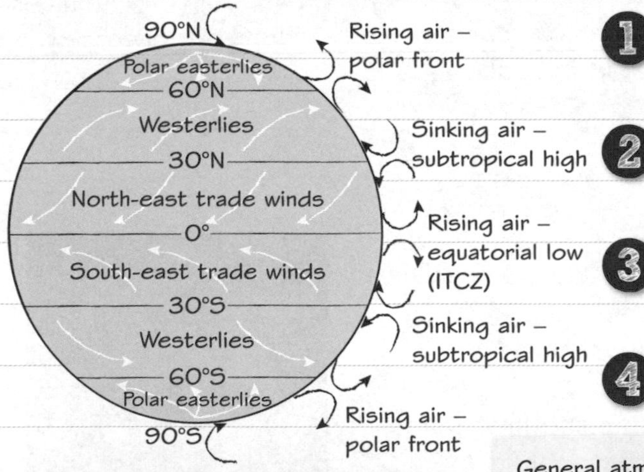

General atmospheric circulation model

1. Winds move outwards from belts of high pressure. Air is sinking here, 'pushing' the surface winds outwards.
2. Winds move towards belts of low pressure. Air is rising here, 'pulling' the surface winds towards the low pressure.
3. The Earth spins as it orbits the Sun, which makes the surface winds curve in a complicated way – known as the **Coriolis effect**.
4. Surface winds transfer huge amounts of heat energy from the Sun round the Earth's surface.

Worked example

Complete this sentence.

At the Equator, hot, rising air creates a belt of _low_ pressure. As air rises, it can retain less water. As a result, the Equator experiences _high_ levels of rainfall.

(2 marks)

Questions like this may seem simple but do not rush them – it could cost you marks.

Now try this

Most tropical storms form between 5° north and 5° south of the Equator. In what direction do the surface winds in this belt blow?

☐ **A** To the east ☐ **B** To the west ☐ **C** To the south-east ☐ **D** To the north-east **(1 mark)**

Tropical storms: distribution

Violent **tropical storms** are known by different names around the world – hurricanes, cyclones and typhoons. You need to know where they occur and how this relates to atmospheric circulation.

Distribution pattern for tropical storms

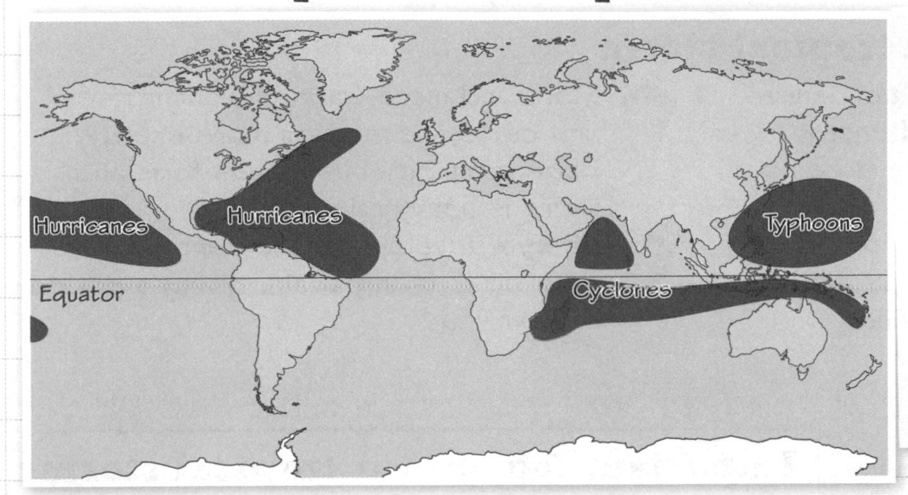

Tropical storms only form where seawater temperatures are above 26.5 °C. This limits:

- their geographical distribution (they occur in the tropics, starting between 5° and 30° of latitude)
- their seasonal distribution (they happen in summer and late autumn when seawater is warmest).

Tropical storms and general atmospheric circulation

1. Tropical storms form in the tropics, near (but not usually at) the Equator, in the permanent belt of low pressure formed by the rising air of the equatorial low. Over the oceans, this air is very moist and warm. It is this moist, warm air that powers tropical storms.
2. Tropical storms have their **source** in the low-pressure belts extending from the Equator, and then **track** westwards. This is because they are affected by surface 'trade' winds, moving from high pressure to low pressure at the Equator, curved westwards by the spin of the Earth.
3.  Some Northern Hemisphere tropical storms reach a belt of surface winds that blow towards the east. This makes the tropical storms change direction, giving them a curved track.

Why are there no tropical storms at the Equator?

As well as needing very low pressure, tropical storms also need some of the Earth's spin to form. The Equator does not have enough spin. The spin begins to have an effect at around 5° north and south of the Equator.

Worked example

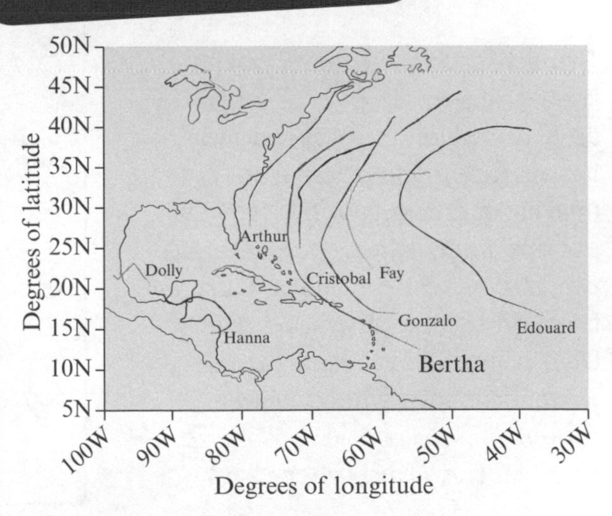

The chart above shows tropical storm tracks in the North Atlantic for July–September 2014. Using the chart, state the approximate latitude and longitude of the source of tropical storm Bertha. **(1 mark)**

12° N, 54° W

Any exam questions can test your geographical skills, knowledge and understanding.

Now try this

Give **one** reason why the hurricanes (tropical storms) shown in the Worked example above first track westwards and then curve eastwards. **(2 marks)**

Had a look ☐ Nearly there ☐ Nailed it! ☐

Tropical storms: causes and structure

Tropical storms require particular conditions to form, and their formation and development occur in a particular sequence.

Five main stages of a tropical storm

1. Warm, moist air rises and condenses, releasing huge amounts of energy.
2. As the air rises, it sucks in more warm, moist air behind it.
3. The air spirals upwards rapidly, intensifying the low pressure and pulling in surface winds. Wind speeds begin to increase.
4. More and more warm, moist air rises and cools, generating more energy. Huge cumulonimbus clouds form and there is heavy rain.
5. An **eye** develops at the centre where drier air is falling again, giving a period of calm.

Worked example

Explain **two** reasons why tropical storms do not form near the UK.

(4 marks)

Tropical storms require water temperatures of over 26.5°C to form, which is a lot warmer than UK sea temperatures (maximum around 17°C). Tropical storms need wind speeds and directions to be constant up through the atmosphere. The UK's climate is dominated by weather fronts where wind speeds and directions change through the atmosphere.

Important factors for tropical storms

- ✓ Warm sea temperatures
- ✓ High humidity – there needs to be a lot of moisture in the atmosphere
- ✓ Rapid cooling – rising air must condense quickly to generate the huge amounts of energy required to power a tropical storm
- ✓ Low wind shear – if winds are blowing in different directions up through the atmosphere, the storm won't form
- ✓ Coriolis force to give the storm spin – this is not usually strong enough within 5° latitude of the Equator
- ✓ Pre-existing low-pressure disturbances – tropical storms usually form when smaller storms come together

A tropical storm

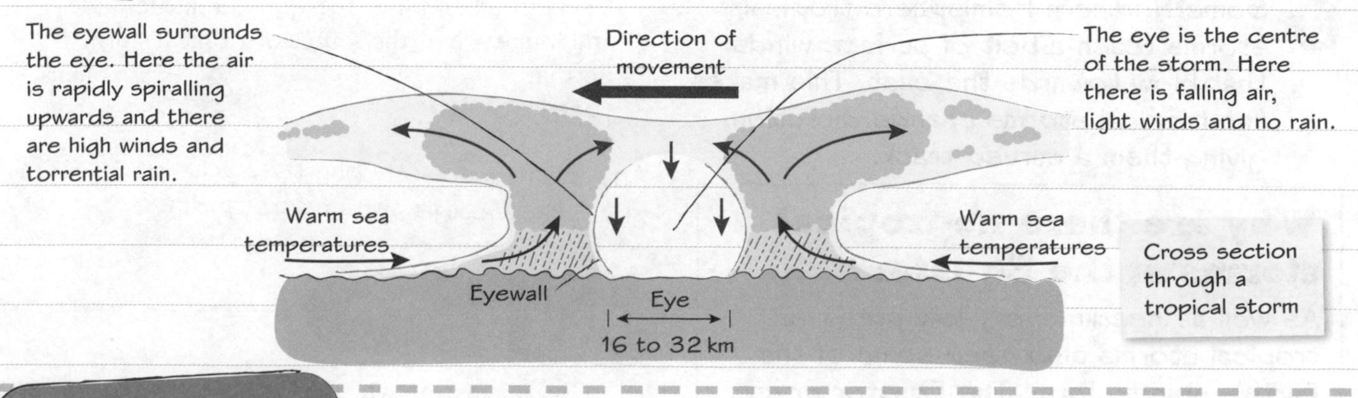

Cross section through a tropical storm

Now try this

Draw your own cross section of a tropical storm's structure, like the one above.
Add numbers 1–5 to show where the five main stages of tropical storm development occur. **(4 marks)**

Tropical storms: changes

Tropical storms vary in **intensity**, and intensity changes as the storm develops. There is evidence that **climate change** is influencing tropical storm intensity, **frequency** (how often they develop) and **distribution** (where they can occur).

The Saffir-Simpson scale

The Saffir-Simpson scale classifies tropical storms into five categories.

Category	Max. wind speed (km/h)	Pressure (millibars)	Storm surge (metres)	Damage
1	119–153	980 and over	1.0–1.7	Some damage – trees lose branches, power lines come down
2	154–177	979–965	1.8–2.6	Roofs and windows damaged, some trees blown over, coastal flooding
3	178–208	964–945	2.7–3.8	Structural damage to buildings. Flooding over 1 m up to 10 km inland
4	209–251	944–920	3.9–5.6	Major devastation – destroys buildings, floods up to 10 km inland
5	252 or higher	<920	>5.7	Catastrophic – destruction up to 5 m above sea level. Mass evacuation needed

What makes tropical storms intensify and dissipate?

Tropical storms **intensify** when:

- water temperatures are over 26.5 °C
- there is low wind shear
- there is high humidity.

Tropical storms **dissipate** (get weaker) when they:

- reach land because they lose energy (they are powered by warm water)
- move into areas of colder water
- meet weather systems where winds are blowing in different directions.

Effects of climate change

- ✓ Tropical storm **intensity** increases when sea temperatures exceed 30 °C. Scientists expect a 2–11 per cent increase this century due to global warming, with stronger winds and heavier rainfall.
- ✓ Tropical storm **distribution** is limited by seawater temperature – most form within 5–15° north and south of the Equator. If sea temperatures increase in the subtropics, this zone could widen to the north and south.
- ✓ The **frequency** of category 4 and 5 hurricanes has increased in the North Atlantic, but overall the frequency may reduce.

Worked example

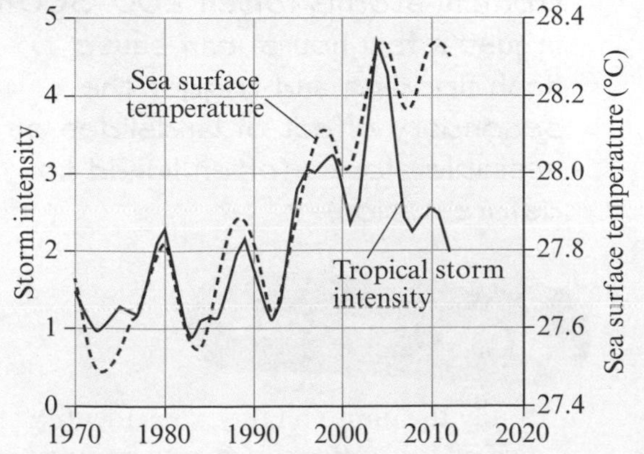

Storm intensity (0–5); Sea surface temperature (°C) (27.4–28.4); 1970 1980 1990 2000 2010 2020

Sea surface temperature

Tropical storm intensity

Study the graph, which plots sea surface temperature and tropical storm intensity. Describe the relationship between sea temperature and tropical storm intensity. **(2 marks)**

From the 1970s there is a close relationship because when sea temperature increases or decreases, so does tropical storm intensity. But in 2009 and 2010, tropical storms became much less intense even though sea temperature remained high.

Had a look ☐ Nearly there ☐ Nailed it! ☐

Tropical storms: effects

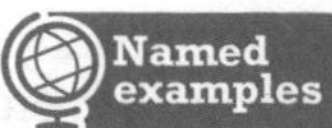

You need to know about the **primary effects** and **secondary effects** of tropical storms, and use a **named example** of a tropical storm to show its primary and secondary effects. This will usually be the same example as for immediate and long-term responses. See page 13 for more about responses.

Primary and secondary effects

- Tropical storms can be very destructive. The **primary effects** are the immediate impacts of the storm itself: strong winds, heavy rain and **storm surges**.
- The **secondary effects** are the longer-term impacts that happen **as a result of** the primary effects. For example, landslides caused by heavy rain, homelessness caused by storm-surge damage, loss of income due to coastal flooding of farmland or tourist resorts.

Worked example

Describe the primary effects of a tropical storm that you have studied. **(4 marks)**

Typhoon Haiyan: a category 5 tropical storm in the Philippines in November 2013. Over 6300 people were killed, most of them drowned by a 5 m storm surge in the city of Tacloban. Over 1 million homes were damaged by the storm surge and the 275 km/h winds. Power lines were blown down, so electricity was cut off. Trees were blown over, blocking roads. As well as coastal flooding from the storm surge, 400 mm of rain caused floods inland, which destroyed crops.

Primary effect: storm surges

- As a tropical storm moves towards a coast, the sea gets shallower. Water pushed up by the wind in front of the storm has nowhere to go but up and onto the land.
- The low atmospheric pressure in the tropical storm also increases the surge.
- Storm surges drown people, sweep away buildings, destroy crops, contaminate agricultural land with salt and cause sewage spills that threaten public health through waterborne diseases, such as typhoid.

The storm surge caused by Typhoon Haiyan in November 2013 reached heights of over 6 metres.

Worked example

Describe the secondary effects of a tropical storm that you have studied. **(4 marks)**

Typhoon Haiyan: a category 5 tropical storm in the Philippines in November 2013. Around 600 000 people had to leave their homes, especially in Tacloban, 90 per cent of which was destroyed. About 400 000 people were housed in 1100 emergency camps. In total, 10 000 schools were destroyed, so children missed out on education. Farmers could not grow rice in soil contaminated by salt and rice prices rose by 12 per cent. There were outbreaks of disease because of contamination by sewage =and because hospitals had been destroyed.

Secondary effect: landslides

Heavy, intense rainfall during tropical storms (often 200–300 mm in just a few hours) can cause flash flooding and trigger the secondary effect of landslides on unstable slopes (often linked to deforestation).

Now try this

Study the photo on this page. Identify **two** primary effects of the storm surge that hit the area shown. **(2 marks)**

Tropical storms: responses

You need to know about the **immediate** and **long-term responses** to tropical storms, and use a **named example** to talk about these responses. This will usually be the **same example** as for primary and secondary effects. See page 12 for the primary and secondary effects of tropical storms.

Immediate responses

- As tropical storms are carefully monitored, **immediate responses** can start **before** the storm makes landfall. For example, organising evacuation of at-risk areas.
- Immediate responses can also take place **during** the tropical storm. For example, rescuing people cut off by flooding.
- Immediate responses can also take place **immediately after** the storm. For example, providing temporary water supplies, removing dead bodies.

Worked example

Describe immediate responses to a tropical storm that you have studied. **(4 marks)**

In August 2005, category 3 Hurricane Katrina caused 1836 deaths in Louisiana, USA, mostly in the city of New Orleans. Although the storm was monitored and its wind speeds and likely points of landfall were accurately predicted, there were other problems. Despite the evacuation process, 60 000 people stayed in New Orleans because they could not leave their homes or did not want to. The Coast Guard rescued 35 000 people from flooded homes. There were too few emergency shelters – 30 000 people arrived at the Louisiana Superdome, which only had enough room for 800. The US Government gave US$60 billion to help survivors, but had problems supplying temporary shelters for the 700 000 residents who needed them.

Worked example

Describe long-term responses to a tropical storm that you have studied. **(4 marks)**

The high number of deaths caused by Hurricane Katrina (August 2005) resulted from the failure of New Orleans' levée system, which had not been properly maintained. Long-term responses to Katrina included making the city's 400 km of levées much higher and much stronger. The city's 78 floodwater pumping stations have all been made flood-proof. The Lake Borgne surge barrier has been built to protect New Orleans – it is currently the largest storm surge barrier in the world. Upgrading these defences cost the US Government US$14 billion.

New funding has been spent on search and rescue teams, and city residents now get evacuation updates by text message.

Long-term responses

- Long-term responses involve repairing and rebuilding areas damaged by the tropical storm.
- Long-term responses also include changes that are made to reduce the effect of tropical storms in the future.

New housing in New Orleans

Now try this

Study the photo above of a new home being constructed in the Lower 9th Ward of New Orleans – an area affected by Hurricane Katrina. Outline **one** reason why this is a long-term response in a tropical storm. **(2 marks)**

Tropical storms: reducing risks

Monitoring, prediction, protection and planning can reduce the effects of tropical storms.

Monitoring and prediction

Satellites, aircraft and powerful computers are at the heart of monitoring and predicting tropical storms.

- **Satellite** images provide data on cloud cover and formations, and allow tropical storms to be tracked over time.
- A **radar** satellite enables tropical storms to be 'scanned' for rainfall; measurements have proved useful in forecasting intensity.
- Planes and drones can fly through tropical storms. **Dropsondes** are dropped through the storm, sending data to nearby aircraft.
- Supercomputers crunch all this data and allow **forecasts** (predictions) to be made. Forecasts can indicate time and location of landfall and likely intensity days ahead of the storm, so evacuations can be organised and preparations made.

Worked example

A meteorologist tracks a tropical storm at the National Hurricane Centre in Miami, USA

Satellite images are useful for identifying tropical storms as they form and also for monitoring their movements. Which **one** of the following is the correct term for this movement? **(1 mark)**

☐ **A** Path ☒ **B** Track ☐ **C** Route ☐ **D** Area

Protection

- People living in wealthy areas in tropical storm zones invest in protection such as storm window shutters or a basement shelter with an emergency generator.
- Governments may invest in protection for at-risk areas, such as high sea walls against storm surges and shelters on higher land.
- Poorer countries can share monitoring and prediction data from wealthier countries, and invest in communication technology to warn communities of a weather hazard.
- Mangrove swamps and coastal marshlands offer natural protection from tropical storms and storm surges. Where possible, governments may protect these environments by stopping development.

Planning

The most effective planning educates people so they know what to do.

WHAT SHOULD I DO?

ARE YOU UNDER THREAT?
Find out from your local Emergency Services whether you are in an area at risk of storm surges. Work out how to get to your nearest shelter safely.

ARE YOU READY TO EVACUATE?
Plan what you will do if you have to evacuate. Will you have essential medicines? Vital documents? What about your pets? Discuss your plans with your local council.

TIME TO EVACUATE!
Be prepared to evacuate as soon as you are advised to. This makes it easier for Emergency Services to manage the difficult task of moving a lot of people all at once, especially if the weather is getting worse. When a storm threat develops, keep listening to official warnings of high tides and coastal flooding from the Bureau of Meteorology.

Now try this

Read the 'What should I do?' text above, which is from the Australian government. What problems regarding dealing with a tropical storm is the government trying to avoid? Suggest **two**. **(2 marks)**

UK: weather hazards

The UK is affected by a number of weather hazards: storms, droughts, heavy and prolonged rainfall and snowfall, heatwaves and severe cold. But rarely all at once!

Heavy rainfall often causes flooding

Heavy rainfall and flooding

The UK is affected by different air masses, giving it very varied weather. **Depressions** coming across the Atlantic bring plenty of rain to the UK, but when depressions combine with **tropical maritime** air, very heavy rainfall can persist for a long time. There was heavy rainfall in December 2015 for these reasons.

Whole days of very heavy rain will lead to flooding. Intense rainfall associated with storm events can also cause **flash flooding**. This occurs when the rain is too intense to be absorbed by the ground and, instead, runs over it.

Winter storms

The UK experiences frequent stormy weather in winter. Some **winter storms** have heavy rain and extremely strong winds. These extreme storms can blow down power lines so that large numbers of people can be without electricity for several days.

When strong winds, low pressure and high tides all occur together at the coast, **coastal flooding** can take place, which can strip beaches of sand and damage homes and transport infrastructure. People are sometimes killed in these extreme winter storms.

Damage to the Exeter–Plymouth railway due to winter storms in 2014

Worked example

In 2010, 2011 and 2012, south, east and central England experienced several dry months. **Reservoir** levels and river levels became very low. Crop and livestock farmers were badly affected by water restrictions. **Hosepipe bans** were introduced to stop people wasting water on washing cars and watering gardens. Wildfires broke out in parts of Surrey, South Wales and Scotland.

Identify **two** negative impacts of droughts, using the text above or your own knowledge. **(2 marks)**

First impact: Farmers may not have enough water for their crops or livestock.

Second impact: There may be wildfires that can destroy houses and kill people.

'First impact' and 'second impact' are already written out for this answer. The exam paper will often do this to save you time.

Now try this

Complete the gaps in the following sentence.

Other examples of extreme weather events in the UK include ______________ with lightning strikes and heavy rainfall and ______________ in which prolonged high temperatures are dangerous for elderly people and people with health problems. **(2 marks)**

UK: extreme weather

You need to know about a recent extreme weather event in the UK: its causes, impacts and how **management strategies** can reduce risk. **Make sure you revise the example you did in class.**

Causes

Extreme weather in the UK is when normal weather becomes intensified.

- Depressions bring wet and windy weather; intense depressions bring heavy rain and strong winds.
- Normally, rain soaks into the ground (infiltration) and drains into rivers, going back to the sea. Extreme rain saturates the ground so no more can soak in. The runoff can cause floods or dumps water into rivers so quickly that they flood.
- High-pressure systems bring clear skies. If a high-pressure system becomes 'blocked', the UK can experience long periods without rain, leading to droughts.

Impacts of extreme weather

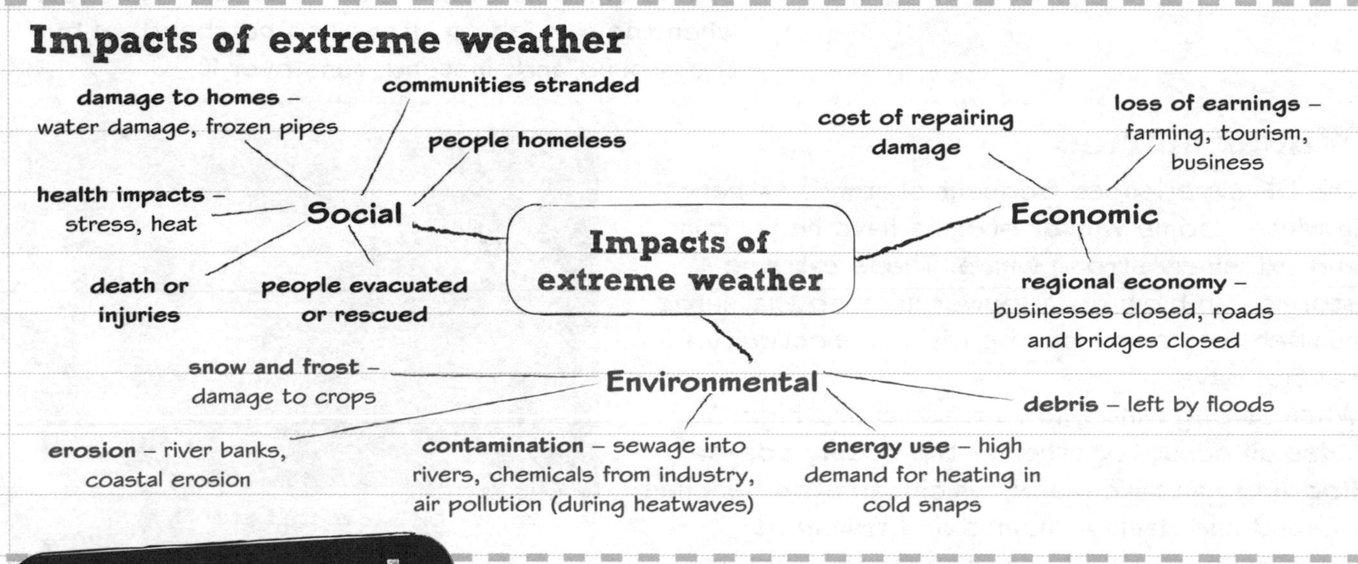

Worked example

Identify the different impacts of an extreme UK weather event you have studied. **(6 marks)**

The floods in Cumbria from 17 to 20 November 2009 (after very heavy rain) had serious social impacts: a police officer was killed after a bridge collapsed, 1500 homes were flooded, ruining people's belongings and causing emotional distress. The floodwater was contaminated with sewage, putting people's health at risk. Economic impacts included the costs of repairing damaged buildings and infrastructure, the insurance costs of replacing and repairing people's damaged property and the costs of new flood defences (£4.5 million). The destruction of six local bridges also had serious economic impacts. People took longer to get to work and it was hard for businesses to move supplies in and out of the area. Environmental impacts included landslides along the banks of the River Derwent and pollution of floodwater with sewage and the destruction of trees pulled down by the floodwaters.

Management strategies

Monitoring and prediction, protection and planning can reduce the negative impacts.

- UK weather forecasters are responsible for identifying extreme weather threats.
- The Environment Agency keeps people informed and can organise evacuations.
- Flood defences and other protection are kept in good repair.
- Plans may include ways to increase resilience to extreme weather.

Now try this

Suggest **one** way in which management strategies can reduce the risk of extreme weather in the UK. **(2 marks)**

UK: more extreme weather

We know that extreme weather events in the UK have impacts on human activity. What is the evidence that weather is becoming **more extreme** in the UK?

Timeline

2004 Flash flooding in Boscastle after extremely heavy rain

2005 Flooding in Carlisle and North Yorkshire

2006 Drought in southern and eastern England

2007 Flooding, especially in Gloucestershire, after record July rain levels

2008 Record low temperature for October

2010 'Big freeze' in January

2012 Wettest year in England since records began

2013–2014 Extreme winter storms: wind damage, river and coastal flooding

2015 Heavy rain and flooding in the north following Storm Desmond

Evidence for more extreme weather

- There has been an increase in the amount of **winter rainfall** and in **winter flooding** since the 1980s.
- **Winter temperatures** in the UK have risen in recent years. In December 2015, the average UK temperature was 8°C (4°C higher than usual). However, December 2010 was one of the **coldest Decembers** on record.
- The UK's annual rainfall totals have **not increased overall**, just the winter amounts.

Is climate change a factor?

The UK's weather is very complex and there have always been extreme weather events. It is not possible to say that global warming is causing the UK to have more extreme rainfall and flooding events. However, in theory, global warming would increase weather extremes because:

- warmer oceans produce more water vapour
- warmer air can hold more water vapour, which means more rain
- warmer temperatures provide more energy for storms, making them more intense.

Worked example

Study this map, which shows rainfall in the winter of 2015–16 as a percentage of average rainfall between 1981 and 2010 in the UK. Describe the distribution of rainfall shown. **(2 marks)**

The map shows that large parts of the north of the UK received between 50 and 150 per cent more rainfall than average. Rainfall was lower in central and southern England, but still often between 10 and 30 per cent more than average. In the east, rainfall was generally lower than average.

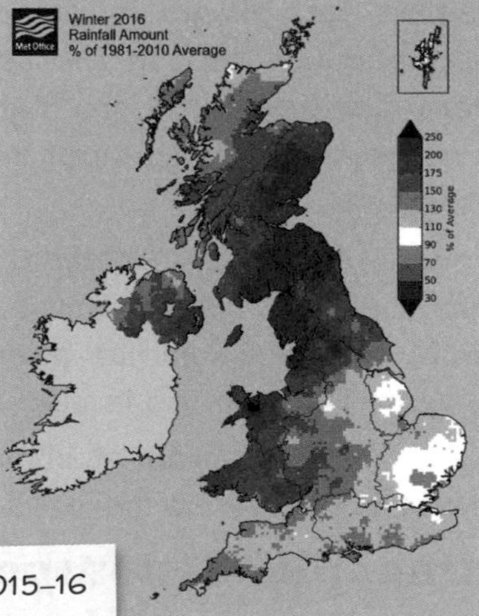

UK rainfall, winter 2015–16

Now try this

A geographer is looking for evidence that extreme weather is becoming more frequent in the UK. Which **one** of the following statements best describes how useful the map above would be for this geographer?

☐ **A** Very useful, as it shows that rain is significantly above average in many regions

☐ **B** Quite useful, as it indicates that regions are receiving rainfall above and below the average

☐ **C** Not very useful, as it only shows totals for one winter **(1 mark)**

Climate change: evidence

There is a range of sources that provide **evidence** for **climate change.** This evidence shows that though climate has often changed in the past, recent changes appear to be much more rapid.

Temperature changes

This graph shows how the Earth's temperature has cooled and warmed over 450 000 years. It demonstrates **long-term** temperature changes due to natural causes.

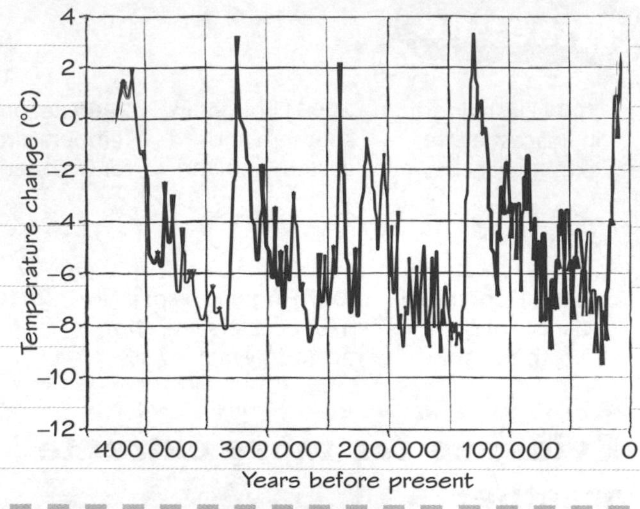

The data for this graph comes from ice cores and fossils. **Ice cores** are cylinders of ice obtained by drilling through glaciers up to 3 km deep in ice up to 500 000 years old.

Sea level rise

Over the last 100 years, global sea levels have risen by 170 mm, and the rate of sea level rise is increasing. This is **evidence of climate change** because:

- seawater expands as it gets warmer
- as ice on the land melts in warmer temperatures, more water enters the sea.

Glacial retreat and Arctic sea ice minimum

Glaciers get bigger in longer winters as more snow accumulates. In shorter winters, and if yearly temperatures are higher, glaciers start to retreat. Globally, most glaciers have been retreating.

There is a minimum extent of Arctic sea ice before winter sets in and it starts to increase again. This minimum has been decreasing year to year. The sea ice is also thinning.

These are **evidence of climate change** because: ice melts more when it is warmer.

Atmospheric temperature rise

Atmospheric temperature measurements show an increase of around 0.8°C between 1880 and 2012. The last 30 years have been the warmest in 1400 years. In total, 16 of the 17 warmest years ever recorded have occurred since 2000.

This is **evidence of climate change** because: heat drives the world's climates.

Worked example

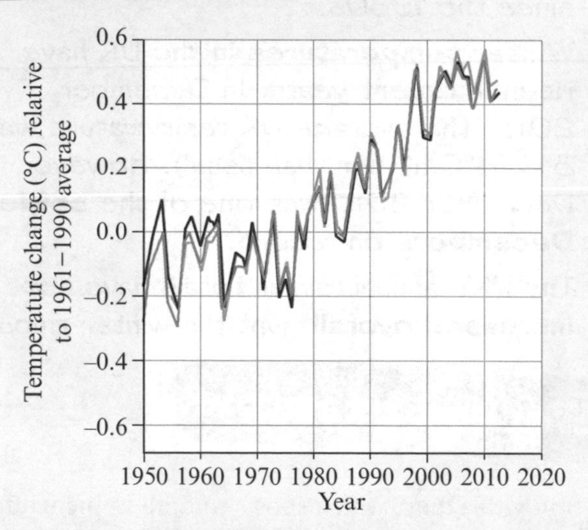

Which **one** of the following is measured by this graph? **(1 mark)**

☐ **A** Extent of minimum Arctic sea ice
☒ **B** Average annual atmospheric temperature
☐ **C** Average extent of Northern Hemisphere glacial retreat (km)
☐ **D** Atmospheric carbon dioxide (ppm)

Check the units of the graph's vertical and horizontal axes.

Now try this

Explain **one** way in which scientists are able to measure what atmospheric temperatures were 400 000 years ago. **(2 marks)**

Climate change: possible causes

Past climate change was due to natural factors, including orbital changes, volcanic activity and changes in solar output. Present climate change is closely linked to human factors.

Some natural causes of climate change

- The Earth's orbit changes a small amount once every 100 000 years. These are known as **Milankovitch cycles.**
- The amount of energy radiated from the Sun changes over an 11-year cycle.
- Volcanic eruptions pump ash dust into the atmosphere causing a cooling effect.
- Large asteroid collisions can cause cooling as material blocks out the Sun. Meteorites hitting the Earth can cause huge fires releasing massive amounts of CO_2 which subsequently has a warming effect.
- Ocean current changes can cause cooling and warming. In the UK, we have a warm and wet climate because of warm Atlantic currents. Sometimes the current shifts and we get a cooler climate for a period of time.

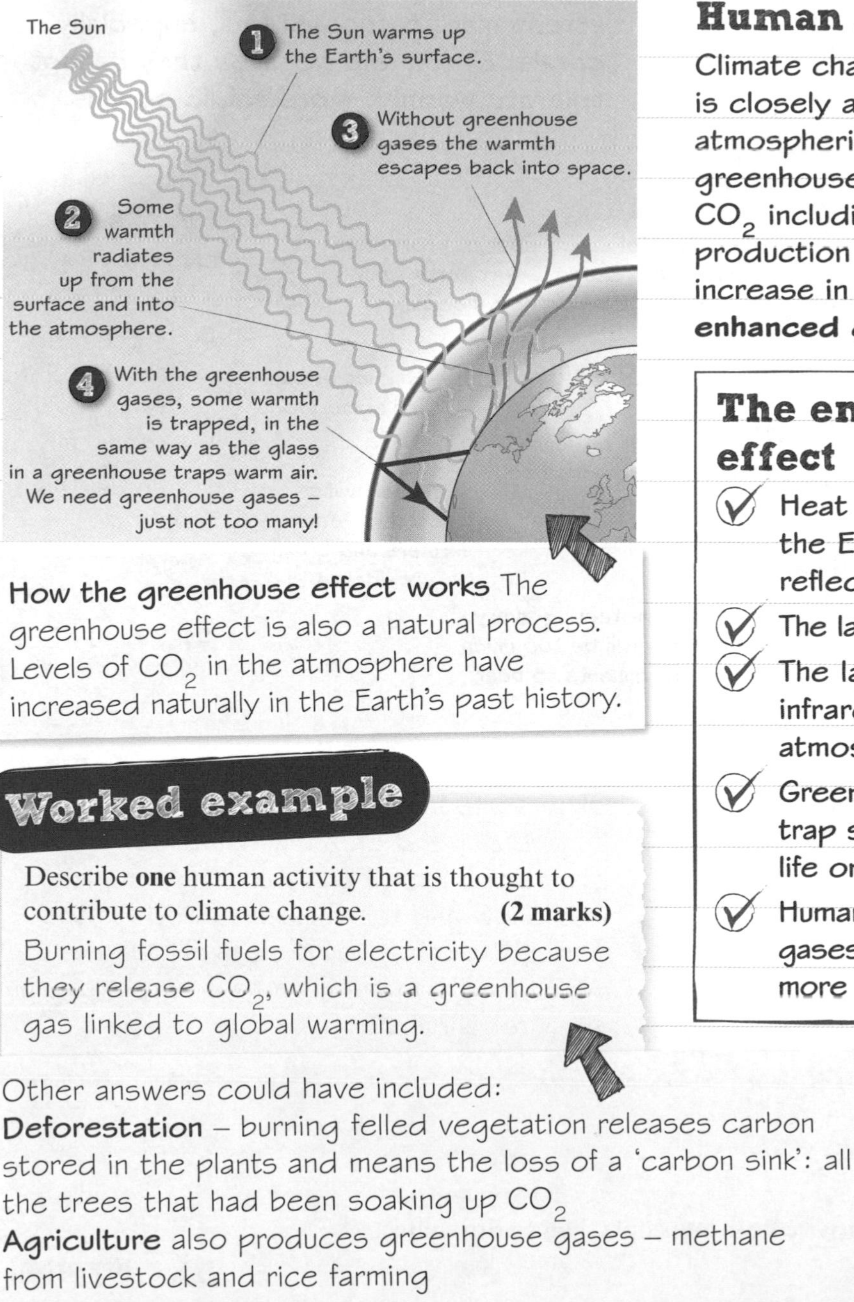

How the greenhouse effect works The greenhouse effect is also a natural process. Levels of CO_2 in the atmosphere have increased naturally in the Earth's past history.

Worked example

Describe **one** human activity that is thought to contribute to climate change. **(2 marks)**

Burning fossil fuels for electricity because they release CO_2, which is a greenhouse gas linked to global warming.

Other answers could have included:
Deforestation – burning felled vegetation releases carbon stored in the plants and means the loss of a 'carbon sink': all the trees that had been soaking up CO_2
Agriculture also produces greenhouse gases – methane from livestock and rice farming

Human factors in climate change

Climate change resulting from global warming is closely associated with rising levels of atmospheric carbon dioxide. CO_2 is a powerful greenhouse gas. Many human activities release CO_2 including industry, transport, energy production and farming. This human-produced increase in atmospheric CO_2 produces the **enhanced** greenhouse effect.

The enhanced greenhouse effect

- ✓ Heat (UV rays) from the Sun reaches the Earth's atmosphere; some is reflected back into space.
- ✓ The land and oceans absorb the heat.
- ✓ The land and oceans then radiate infrared heat back into the atmosphere.
- ✓ Greenhouses gases in the atmosphere trap some of the heat (necessary for life on Earth!).
- ✓ Human activity increases greenhouse gases in the atmosphere, leading to more warming.

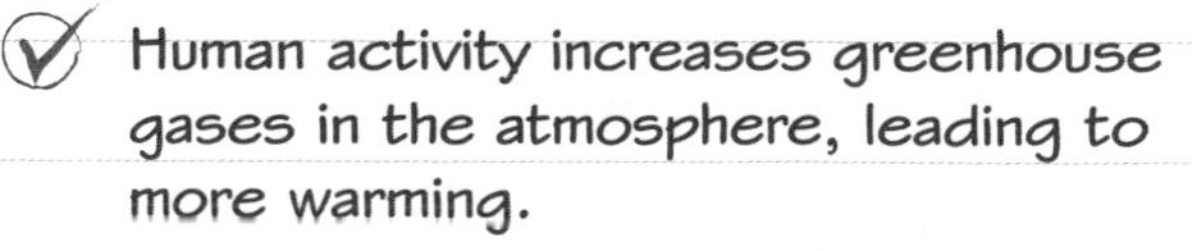

Now try this

Identify **two** of the main greenhouse gases associated with human activity. **(2 marks)**

Climate change: effects

Climate change is already having effects around the world. Future climate change could have major effects on the environment and on people.

Climate change consequences

Rising global temperatures could trigger a chain of negative consequences for people.

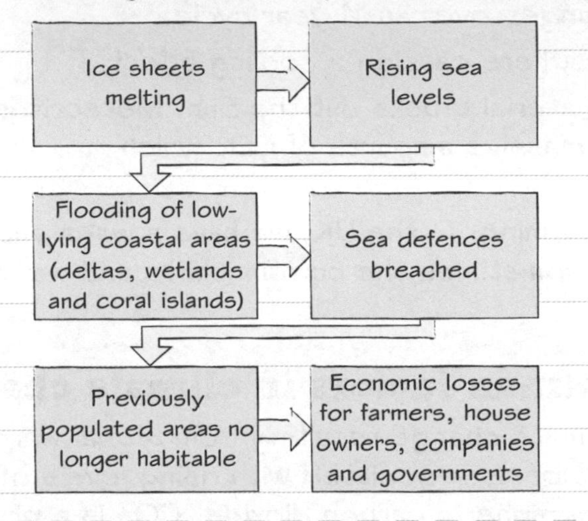

Environmental effects

- Tropical rainforests may become drier and hotter. This would lead to massive loss of biodiversity as many rainforest plant and animal species die out.
- Polar ecosystems would have less ice through the year. This puts stress on animals adapted to ice environments, such as seals and polar bears.
- Desertification would increase if semi-arid areas experienced longer droughts.
- Increases in sea temperature would stress marine ecosystems, especially corals. Corals die because they cannot tolerate warmer, more acidic oceans.

Effects on people

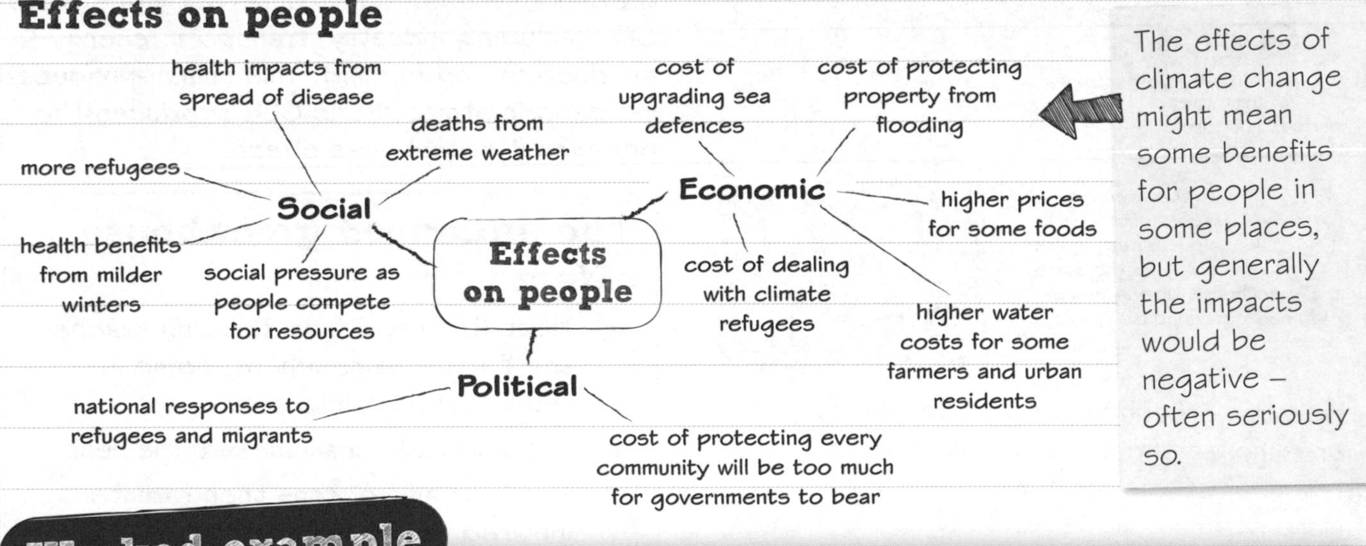

The effects of climate change might mean some benefits for people in some places, but generally the impacts would be negative – often seriously so.

Worked example

Outline **one** possible economic impact of climate change. **(2 marks)**

If a question asks specifically for an economic impact, make sure to write about an impact that has a financial cost.

Ski resorts in North America and in Europe depend on snow to attract winter sports tourists. If climate change means less snow and for shorter periods in the winter, then this would mean ski resorts would have to pay more to create artificial snow for their slopes.

Now try this

'Climate change would have more negative effects on people than positive effects.' Use evidence to support this statement. **(6 marks)**

Mitigating climate change

Strategies to manage climate change include **mitigation**: – reducing the causes of climate change. Mitigation strategies you need to know about are: alternative energy production, carbon capture, planting trees and international agreements.

International agreements

1. **Kyoto Protocol (1997)** – 37 industrialised countries agreed to cut emissions of CO_2 by 2012, but few met their targets.
2. **Copenhagen Accord (2009)** – this new agreement softened the terms of Kyoto. Countries were asked to say what cuts they could make to CO_2 emissions by 2020, but there was no law to make sure these cuts were made.
3. **Paris Agreement (2015)** – 195 countries agreed to act to limit global average temperature rises to below 2°C. Countries submitted national plans to cut emissions from 2020. US$100 billion went to poorer countries to help them develop without increasing emissions. In 2016, China and the USA, which generate 40 per cent of world CO_2 emissions, both agreed to sign up to the agreement. However, Paris is a voluntary agreement: countries that do not put their plans into action will not be punished.

Alternative energy production

👍 Nearly 90 per cent of carbon emissions from human activities are from burning fossil fuels. Alternative energy offers sources that do not emit CO_2:

- wind and solar power
- nuclear power
- hydro-electric power.

👎 Alternative sources are not as effective at providing power as fossil fuels. People find it hard to accept using less power, or paying more for the power they use.

Solar power now supplies 178 gigawatts of electricity – 1 per cent of global demand

Carbon capture and storage (CCS)

👍 Filters can capture carbon dioxide as it is emitted. The gas can be compressed and transported in pipes to natural cavities deep underground: for example, aquifers, worked-out coal or gas fields or porous rock formations.

CO_2 can also be 'scrubbed' out of the atmosphere and stored in the same way.

👎 Fossil fuel power stations could be fitted with CCS. However CCS is costly, and it makes energy production harder: an extra 40 per cent of fossil fuel is needed to get the same results. This makes energy produced by CCS more expensive.

Planting trees

👍 15 per cent of global carbon emissions result from deforestation. Trees absorb carbon dioxide from the atmosphere and release oxygen through photosynthesis. They store the carbon they absorb, releasing it slowly when they die and decompose. Replanting forests is a good mitigation strategy and is cheap. Since 1978, China has planted 66 billion trees.

👎 It takes decades for trees to grow fully, and there is evidence that wild forests are much better at absorbing carbon than human plantations. Unless we tackle high deforestation rates, replanting will have little impact on carbon emissions.

Now try this

Complete this statement.

Mitigating climate change means ______________ the ______________ of climate change. **(2 marks)**

Had a look ☐ Nearly there ☐ Nailed it! ☐

Adapting to climate change

As well as mitigation, strategies to manage climate change also include adaptation.

Changing agricultural systems

Around 80 per cent of the world's agriculture depends on rain, but climate change is likely to bring more drought to many regions. Options include:

- introducing irrigation (though pumping water uses a lot of energy)
- introducing drought-resistant crops
- changing to different crops that need less water, or to livestock farming.

The threat to water supply

Climate change threatens supplies in many ways.

- ✓ More droughts would increase demand for water from farmers.
- ✓ Many rivers are fed by glaciers through the year, so if glaciers disappear so might rivers, at least for some of the year.
- ✓ Intense rainfall does not recharge groundwater when it produces only surface runoff.

Reducing risk from rising sea levels

Options include:

- retreat from at-risk coastal areas or rely on natural defences against coastal flooding (salt marshes, mangroves)
- adapt buildings in at-risk areas to be more resilient: for example, buildings on stilts, buildings on rafts
- construct large-scale flood barriers and sea walls: London's Thames Barrier is one example; St Petersburg (Russia) is planning a £2 billion barrier to protect against sea rises and surges of up to 4 metres.

There are disadvantages to all these adaptations, for example flood barriers are extremely expensive.

Managing water supply

Options include:

- **small-scale** – improving rainfall storage for farmers: for example, by constructing sand dams in river beds (these soak up water through the wet season and store it into the dry season)
- **medium-scale** – introducing new technologies to use less water more efficiently, such as smaller flush toilets and drip-feed irrigation
- **large-scale** – water desalination plants using solar energy to make sea water drinkable. Another method is waste water recycling 'from toilet to tap': cleaning up water used by people so more people can use it again.

Worked example

Study the photo, which shows a community of amphibious (floating) houses in The Netherlands. Suggest **two** other ways in which people could reduce the risk of rising sea levels. **(2 marks)**

Building levées (earth banks along the coastline) is one way to cope with sea level rise. It is cheaper than building with concrete and easier to repair.

A second way is for governments to be stricter about what types of land use are allowed near the coast: for example, not homes.

Now try this

Which **two** of the following are examples of mitigation?

☐ **A** Introducing new irrigation technology to farmers in regions at risk of drought

☐ **B** Planting trees to increase carbon storage

☐ **C** Installing solar power on all government buildings

☐ **D** Repairing leaking water pipes **(2 marks)**

A small-scale UK ecosystem

Example You need to know an **example** of a small-scale UK ecosystem you can use to illustrate the interrelationships within an ecosystem: producers, consumers, decomposers, food chain, food webs and nutrient cycling. **Make sure you revise the example you learned in class.**

A hedgerow: a small-scale ecosystem

The wet climate and rich soils of the UK mean a wide variety of plants grow in a hedge supporting many different organisms.

A hedgerow

home to animals, birds and insects

climate is warm in summer, wet and cool in winter

hedgerow plants

fertile soils enriched by rotting vegetation

Food chains and food webs

A food chain

Producers use energy from the Sun to make food

Consumers get their energy from eating producers or ...

Some consumers eat other consumers to get their energy

In an ecosystem the flow of energy and the cycling of nutrients need to be constant to maintain the balance. Nutrients are recycled with the help of decomposers which break down the dead remains of plants and animals and release chemicals for plants to use again.

A **food chain** describes the flow of energy from a producer to a herbivore (plant consumer) and on to a carnivore (herbivore consumer). At each stage, less energy is passed on.

A **food web** illustrates all the different food chains in an ecosystem and how they connect together. In a small-scale ecosystem such as an oak tree, aphids and caterpillars feed on the leaves, spiders eat aphids, blue tits eat spiders and caterpillars, and sparrowhawks eat blue tits.

Worked example

A simple food chain on this page shows that rabbits eat plants and foxes eat rabbits. Suggest what might happen if there was a sudden change to the number of rabbits.

(4 marks)

If the number of rabbits dropped suddenly, plants such as grasses would grow more. But the number of foxes might drop if rabbits were what they mostly ate. However, if the number of rabbits increased rapidly, there might be a growth in the fox population for a while. But, then the rabbits might eat all the grass, rabbit numbers would crash and then fox numbers would too.

Now try this

What is meant by the term **nutrient cycling**? **(1 mark)**

Questions like this are looking for a definition only.

Ecosystems and change

Ecosystems are made up of different components – plants, animals and their environment – in balance. A significant change in one component can put the whole ecosystem out of balance.

Nutrient cycle diagram of a tropical rainforest ecosystem

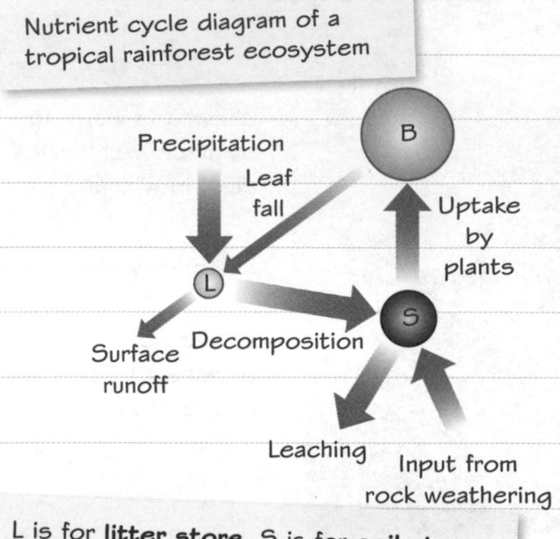

L is for **litter store**, S is for **soil store** and B is for **biomass in nutrient cycles.**

Nutrient cycle: tropical rainforest

A tropical rainforest ecosystem has **inputs** of rain, sunlight and atmospheric gases, and nutrients from the weathering of bedrock. It loses **outputs** through surface runoff and leaching – when minerals are carried out of the soil by water moving down into the bedrock.

- Within the ecosystem, nutrients (energy) are **cycled** between its three stores – biomass, soil and litter. Plants take nutrients from the soil; plants drop their leaves; nutrients return to the soil.
- A change in the ecosystem can unbalance it. For example, if rainfall decreases, decomposition happens less rapidly (decomposition is fastest in warm, wet conditions).

Food web: ancient woodlands

Changes within an ecosystem food web can change the interrelationships within the ecosystem. These changes can be natural or due to human activity.

For example, deer are a part of the New Forest ecosystem (it was created as a deer hunting area by William the Conqueror). However, deer have no natural predators in this ecosystem, so humans have to cull the deer population regularly to stop it growing too large. Deer strip the bark from trees, killing young trees. Too many deer would threaten the future of the forest itself.

Food web for the New Forest ecosystem in southern England

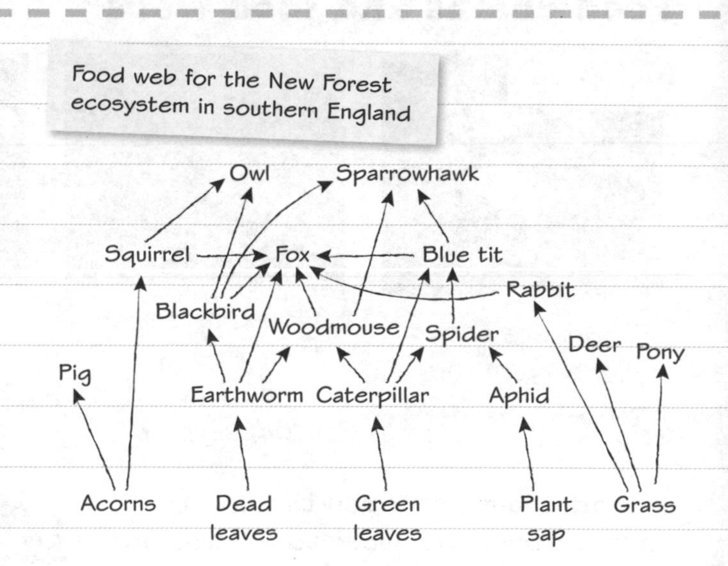

Worked example

Eutrophication from water pollution

The photo opposite shows eutrophication in a pond ecosystem. Complete the following statement. **(2 marks)**

Eutrophication happens when agricultural fertiliser drains into a pond. This increase in the nutrient supply causes an algae 'explosion', which upsets the balance of ecosystem components.

Now try this

Study the rainforest nutrient cycle on this page. Suggest **one** reason why rainforest soils quickly lose their nutrients and become infertile after trees have been removed for farming. **(2 marks)**

Global ecosystems

The distribution and characteristics of large-scale natural global ecosystems are strongly influenced by climate, and climate is driven by the heat energy that different parts of the Earth receive from the Sun. See page 8 for the general atmospheric circulation, which will come in useful with this topic.

Taiga ecosystem

Taiga (boreal) forests are at higher latitudes where the Sun's rays are weak. Trees are adapted to the cold with needle-like leaves.

Temperate ecosystem

Temperate forests have high rainfall and there are seasonal variations in the Sun's rays. Trees lose their leaves in the cool winters.

Tundra ecosystem

The **tundra** is within the Arctic Circle. The Sun gives little heat here and there is little rainfall. Only tough, short grasses survive.

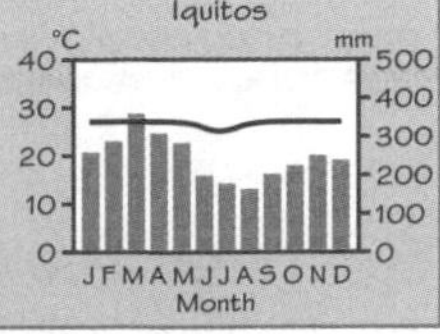

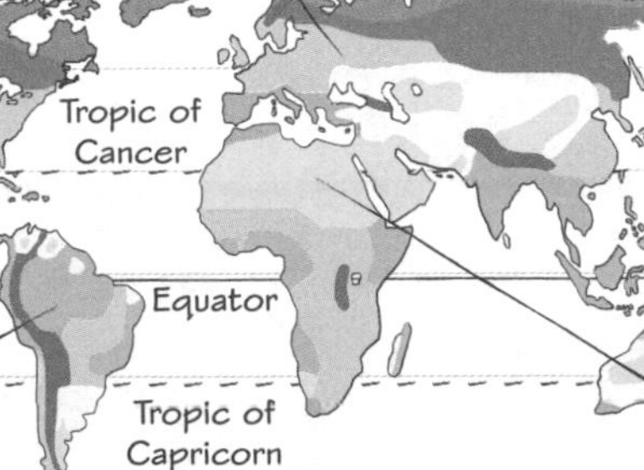

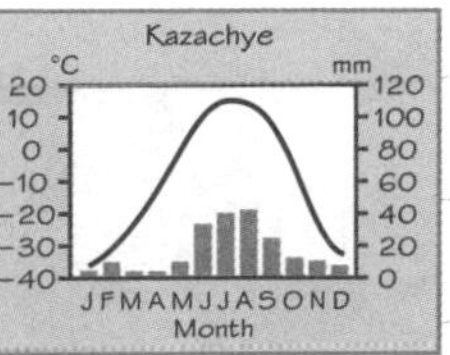

Tropical forest | Tropical grasslands | Desert | Mediterranean | Temperate forest | Mountain | Temperate grassland | Taiga forest | Tundra

Tropical ecosystem

Tropical rainforests are mostly found either side of the Equator. The temperature is hot and there is heavy rainfall.

Desert ecosystem

Deserts are close to the tropics of Cancer and Capricorn. This is where hot, dry air sinks down to the Earth's surface and the Sun's rays are concentrated making it very hot in the day.

Worked example

Explain how climate controls the distribution of tropical rainforest ecosystems. **(4 marks)**

Climate plays an important role in the distribution of all ecosystems. Areas closest to the Equator, where the Sun's rays are most direct, have hot temperatures. Further north or south of the Equator, there is less sunlight and the Sun's rays are weaker, so temperatures are colder. Sunlight, temperature and rainfall affect the type and number of plants that can grow.

Tropical rainforests contain the greatest variety and density of plant growth because they are located in the climate zones where temperatures are warm (near the Equator) and wet (because of the low-pressure zone created by rising warm air at the Equator). Further north and south of the tropical zone, conditions are too dry for tropical rainforest ecosystems because this is where the descending dry air creates high-pressure zones.

This question is not asking for a **description** of the distribution but an **explanation** of something specific: how climate controls the distribution of rainforest ecosystems.

Note how the student has used their knowledge of the general atmospheric circulation model in their answer.

Now try this

1 In which one of the following large-scale global ecosystems is the UK located?

☐ **A** Taiga ecosystem
☐ **B** Tundra ecosystem
☐ **C** Temperate ecosystem
☐ **D** Tropical ecosystem **(1 mark)**

2 The map on this page shows the distribution of the large-scale natural global ecosystems. Describe the distribution of the tundra ecosystem. **(2 marks)**

Rainforest characteristics

Tropical rainforest ecosystems have distinctive characteristics because of the way that the warm, wet tropical climate creates the perfect conditions for massive plant and animal **biodiversity**.

The structure of a tropical rainforest

The rainforest has a distinctive layered structure as trees compete for light.

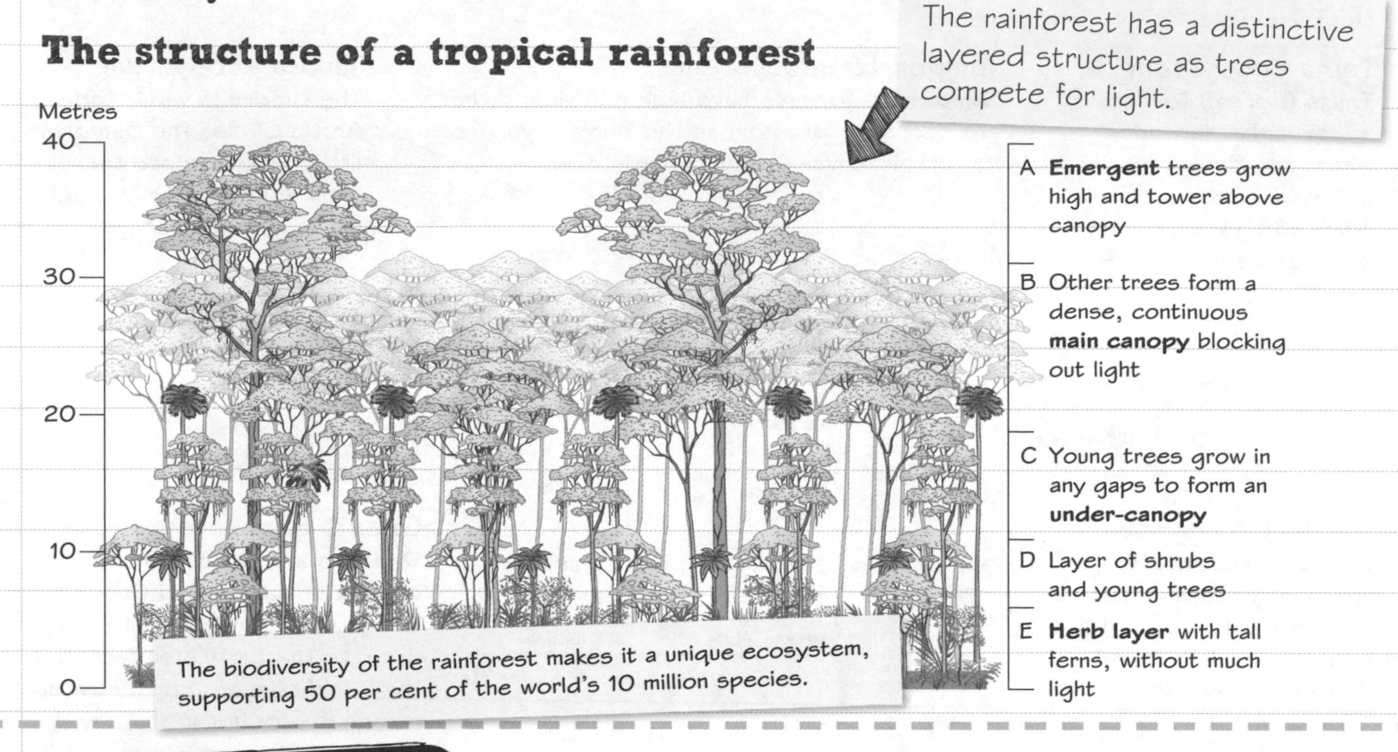

The biodiversity of the rainforest makes it a unique ecosystem, supporting 50 per cent of the world's 10 million species.

Worked example

The diagram below shows a tropical rainforest soil profile. Outline **one** reason why tropical rainforest soils quickly lose nutrients. **(2 marks)**

Tropical rainforest ecosystems have heavy rainfall, with over 300 mm a month in wet seasons. This leaches nutrients out of the soil.

This answer uses knowledge about rainforest climates to add detail to the explanation.

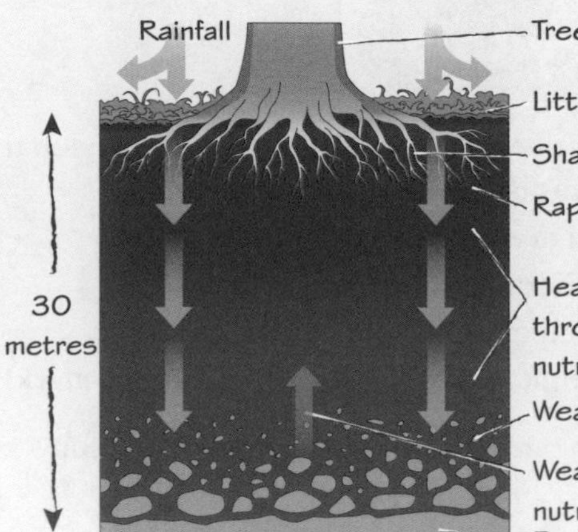

Tropical rainforest soil profile

Tropical rainforest climate

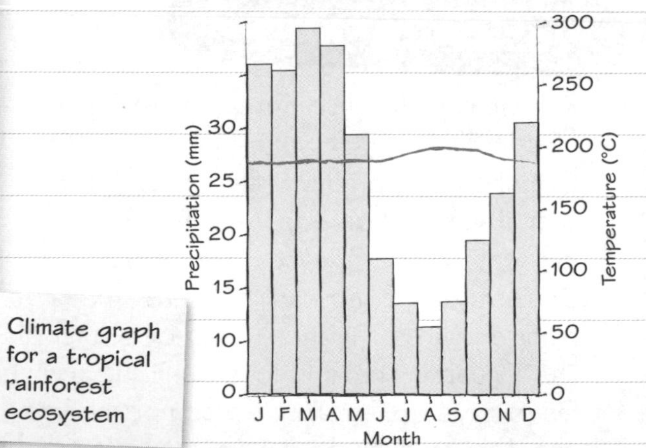

Climate graph for a tropical rainforest ecosystem

Tropical ecosystems have a distinctive climate, with rainfall and warm temperatures all year round. There is no autumn where the trees all lose their leaves. Plants grow constantly, all year round.

Now try this

Complete this statement.

Rainforest trees grow very tall (30–45 m). One other physical characteristic of these trees is that they have ________________.

(1 mark)

Interdependences and adaptations

In the tropical rainforest, climate, water, soils, plants and animals are **interrelated** and **interdependent**. Plants, animals and rainforest peoples have adapted to different ecological niches.

Why is biodiversity so high?

1. Perfect climatic conditions all year round for plant life support huge numbers of species.
2. Large number of plant species provides food for a similarly large number of animal species.
3. Stratification (layers) of the rainforest produces different **ecological niches**. Different species evolve to meet the different needs of these niches.
4. Tropical rainforests are ancient – plants and animals have been evolving there for hundreds of thousands of years.

How ecosystem components interrelate

- Water enters the ecosystem as rainfall – so much rainfall that nutrients in the soil leach out.
- Plants take nutrients and water from the soil rapidly – the constant warmth and wetness are ideal for plant growth.
- Nutrients return to the soil as plants die or lose leaves and animals die and decompose. Nutrients also enter the soil from minerals weathered from the bedrock.

Top three plant adaptations

1. **The dense forest canopy blocks out light**. Some trees, called **emergents**, grow 40 m tall, 10 m above the canopy.

2. **Mould grows on all wet surfaces**. This would block sunlight from leaves. Most plants have evolved **drip tips** that channel water off the leaf.

3. **Nutrients are concentrated in only the top layer of the soil**. This means tree roots have to be shallow. **Buttress roots** give the tall trees extra stability.

Animal adaptations

Animals that live in the trees have strong limbs for jumping and climbing; animals that live on the forest floor are often nocturnal. Birds have short, powerful wings for flying through the dense canopy.

Worked example

Study this photo of a chameleon from a rainforest in Madagascar. Identify **two** ways in which this chameleon is adapted to the rainforest environment. **(2 marks)**

Many animals use camouflage to avoid predators in the rainforest. Chameleons can change their skin colour to match their surroundings. The chameleon has feet that are adapted to grip branches and a tail that it can wind around branches in the canopy.

Two valid ways are identified, with information on why they are adaptations to the rainforest.

Now try this

Tropical rainforest trees often have buttress roots. Outline **one** reason why this is an adaptation to the physical conditions of the tropical rainforest. **(2 marks)**

Deforestation: causes

Case study Each year, 7.3 million hectares of rainforest are cleared: the equivalent of 36 football pitches of forest every minute. You need to be able to use a **case study** of tropical rainforest to talk about the causes of deforestation.

Commercial farming

Commercial farming, when farmers grow crops to sell, is the main cause of deforestation worldwide.

- 80 per cent of rainforest deforestation in Brazil over the last 20 years was for **cattle farming** – beef is a key export.
- More recently, **growing sugar cane** has also led to large-scale deforestation.
- In South East Asia, rapid deforestation has occurred as **palm oil plantations** have replaced forest.

How is commercial farming involved in deforestation in your case study?

Subsistence farming

This is when people grow food to feed their family, clearing plots using 'slash and burn'. After a year or two, they would clear a new plot.

Population growth

Slash and burn clearance for farming used to be sustainable because the rainforest grew back on the old plots. However, rapid population growth has put more pressure on rainforests, with more people clearing land. People keep using their land until all the nutrients are lost and the land will no longer grow anything.

About one-third of rainforest deforestation is the result of population growth and subsistence farming. How important are these in your case study?

Other causes of deforestation

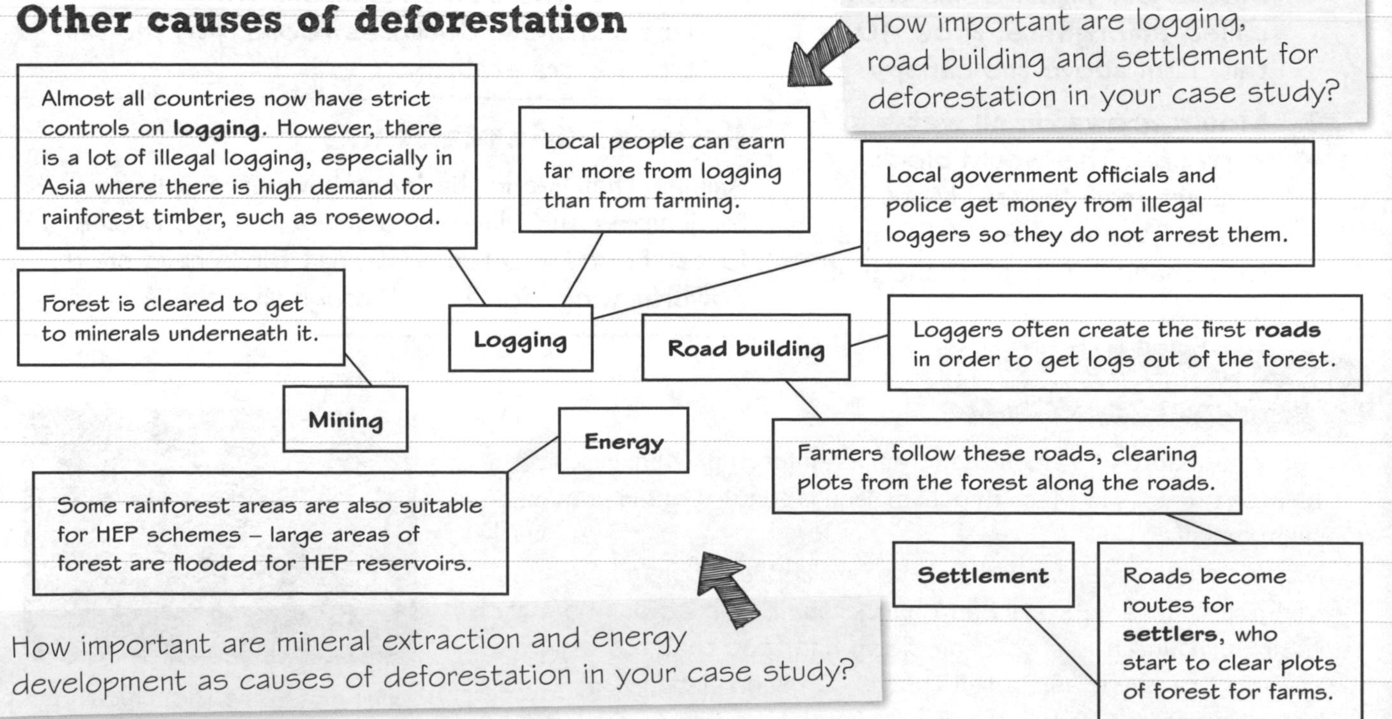

Now try this

Using your own case study information, explain how **two** of the following are involved in causing tropical rainforest deforestation: subsistence and commercial farming, logging, road building, mineral extraction, energy development, settlement and population growth. **(6 marks)**

Deforestation: impacts

Case study As well as using your **case study** to write about causes of deforestation, you also need it to illustrate your answers on the impacts of deforestation: economic development, soil erosion and contribution to climate change.

Worked example

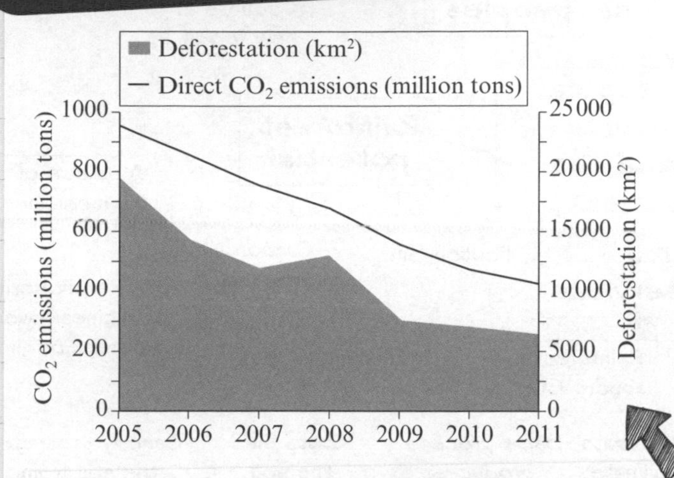

Study the graph on the left, which compares the rate of deforestation in the Brazilian Amazon with CO_2 emissions.

Suggest **one** reason why a reduction in the amount of forest lost to deforestation is closely linked to a reduction in CO_2 emissions. **(2 marks)**

A NASA study showed that tropical rainforests absorb 1.5 billion tonnes of atmospheric CO_2 a year. If deforestation rates slow that means there is more forest to soak up CO_2.

This graph shows how the reduction in the rate of deforestation in the Brazilian Amazon rainforest between 2005 and 2011 was accompanied but a reduction in CO_2 emissions.

Deforestation and economic development

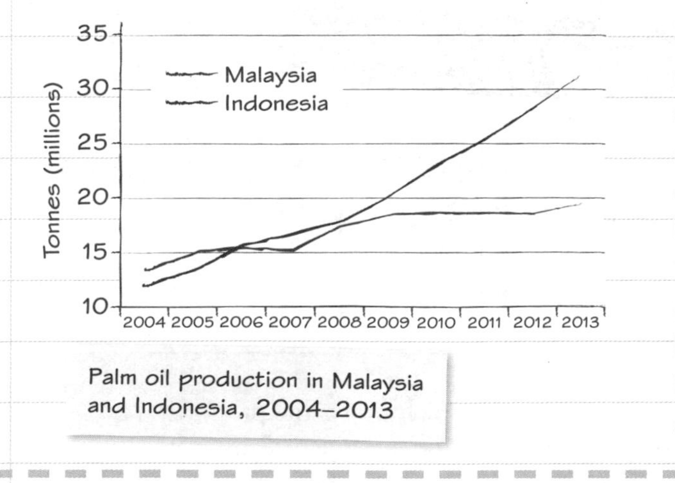

Palm oil production in Malaysia and Indonesia, 2004–2013

Deforestation takes place because the land underneath it has more value than the forest. Countries like Indonesia have used palm oil, grown on deforested land, to fuel their economic development, as Brazil has done with beef, sugar cane and soya crops.

However, the economic benefits of clearing rainforest are often short-lived, because the land declines in fertility so quickly due to leaching. Then the loss of forest begins to be an environmental cost as carbon emissions rise.

Deforestation and soil erosion

Deforestation leaves the tropical soil unprotected from heavy rain, leading to increased surface runoff. A study in Côte d'Ivoire found that, while a hectare of forested slope lost 0.03 tonnes of soil per year to soil erosion, a bare slope lost 138 tonnes of soil. The soil is carried into rivers and causes economic damage, silting up dams and reservoirs. Soil erosion on farmland produces gullies and takes away fertile topsoil.

Now try this

Using your own case study information, explain how **two** of the following are impacts of rainforest deforestation: economic development, soil erosion and contributions to climate change. **(6 marks)**

Rainforest value

Tropical rainforests are valuable to people and to the environment in a wide range of ways. The different ways in which rainforests are valuable influences attempts to manage their use.

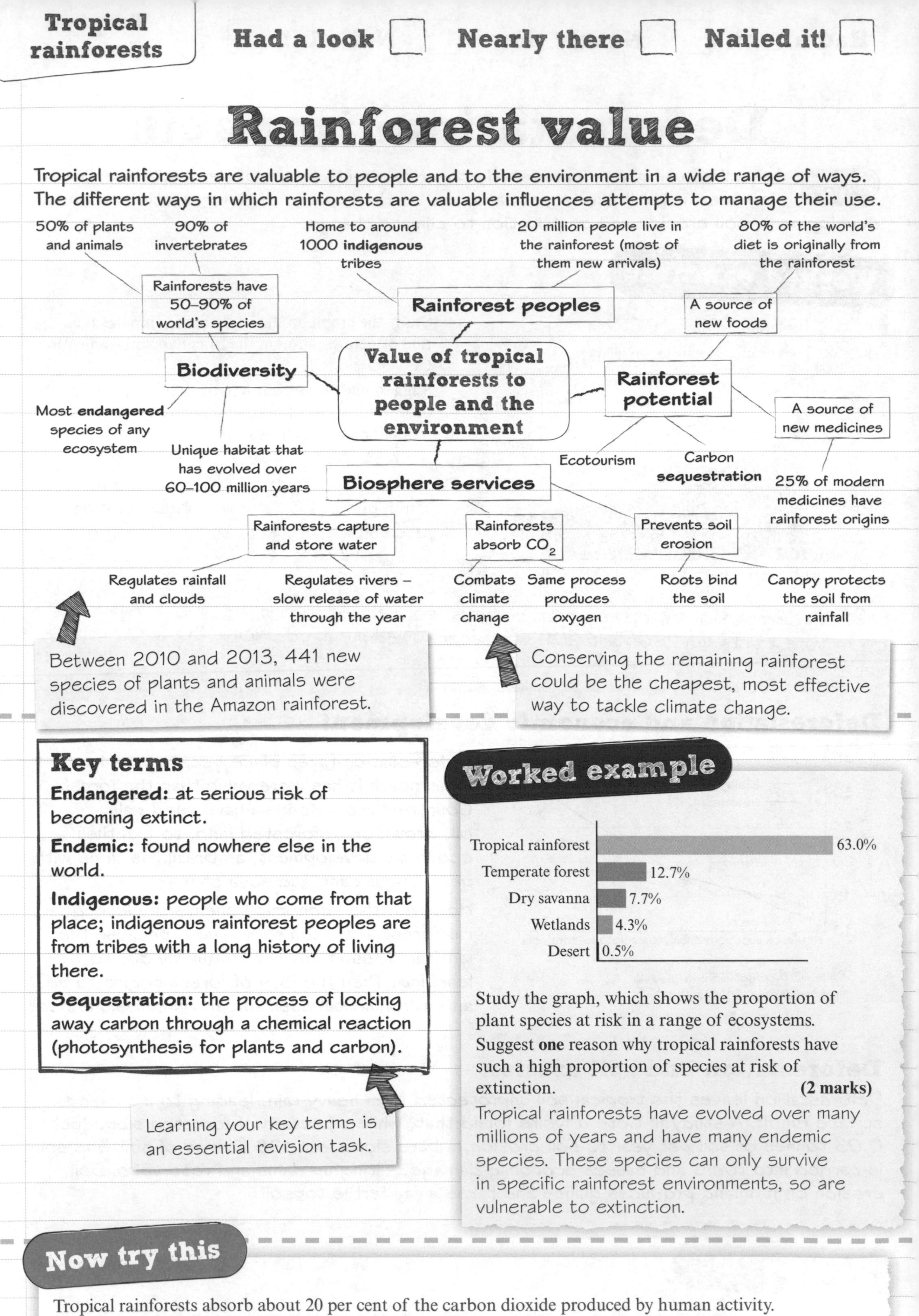

Between 2010 and 2013, 441 new species of plants and animals were discovered in the Amazon rainforest.

Conserving the remaining rainforest could be the cheapest, most effective way to tackle climate change.

Key terms

Endangered: at serious risk of becoming extinct.

Endemic: found nowhere else in the world.

Indigenous: people who come from that place; indigenous rainforest peoples are from tribes with a long history of living there.

Sequestration: the process of locking away carbon through a chemical reaction (photosynthesis for plants and carbon).

Learning your key terms is an essential revision task.

Worked example

Ecosystem	Proportion
Tropical rainforest	63.0%
Temperate forest	12.7%
Dry savanna	7.7%
Wetlands	4.3%
Desert	0.5%

Study the graph, which shows the proportion of plant species at risk in a range of ecosystems.
Suggest **one** reason why tropical rainforests have such a high proportion of species at risk of extinction. **(2 marks)**

Tropical rainforests have evolved over many millions of years and have many endemic species. These species can only survive in specific rainforest environments, so are vulnerable to extinction.

Now try this

Tropical rainforests absorb about 20 per cent of the carbon dioxide produced by human activity.
Which **one** of these terms is used to describe this process?

☐ **A** Photosynthesis ☐ **B** Carbon sequestration ☐ **C** Mitigation ☐ **D** Carbon conservation **(1 mark)**

Sustainable management

Sustainable rainforest management means using rainforest resources now in such a way that future generations will be able to use the same resources to meet their needs.

Selective logging

This is when the only trees cut down are from selected species or when trees reach a certain height.

- 👍 More sustainable than clear-cutting (felling all the forest) because the rest of the rainforest still remains.
- 👎 Weeds and vines can grow where trees are felled and stop other trees growing.

Replanting

- 👍 Replanting cleared land with new rainforest trees helps restore some habitat for animals and protects soils from erosion.
- 👎 Rainforests are highly complex, interrelated ecosystems; many plants need specific insects to pollinate them and without the right decomposers in the litter layer the rainforest nutrient cycle will not function. This all makes it difficult to replace tropical rainforest by replanting.

Sustainable rainforest management

This includes selective logging and replanting, conservation and education, ecotourism and international agreements and debt reduction.

The International Tropical Timber Agreement

The International Tropical Timber Organization (ITTO) is an international organisation that aims to make sure all tropical hardwood that is exported onto the international market comes from sustainably managed sources.

Replanting can successfully connect remaining 'islands' of tropical rainforest with 'tree corridors' that allow rainforest animals to travel from one 'island' to another.

Conservation

Conservation schemes offer different levels of protection from harmful human activities. The strictest conservation areas are off-limits to anyone except authorised researchers.

- 👍 Conservation areas can be managed sustainably, to avoid long-term damage to the ecosystem and help endangered species hold on to the habitat they depend on.
- 👎 The people who desperately need land to farm or make a living from are shut out of conservation areas. Governments can be accused of protecting plants instead of helping poor people to live.

Education

The most valuable education is to train farmers to make better use of already deforested land. For example, using charcoal as fertiliser keeps cleared plots fertile for longer. This reduces the pressure on farmers to clear more rainforest.

Worked example

Explain how ecotourism provides a sustainable way of using tropical rainforests. **(4 marks)**

Ecotourism aims to benefit local people without damaging the forest ecosystem. Tourist numbers are kept low, facilities reuse and recycle as much as possible and local people are employed as staff, guides and educators in a way that values traditional knowledge and skills.

Now try this

Suggest how reducing international debt could help protect rainforest resources. **(4 marks)**

Hot deserts: characteristics

Hot desert ecosystems have a range of distinctive characteristics.

Only revise hot deserts if you studied them in class.

Global distribution of hot deserts

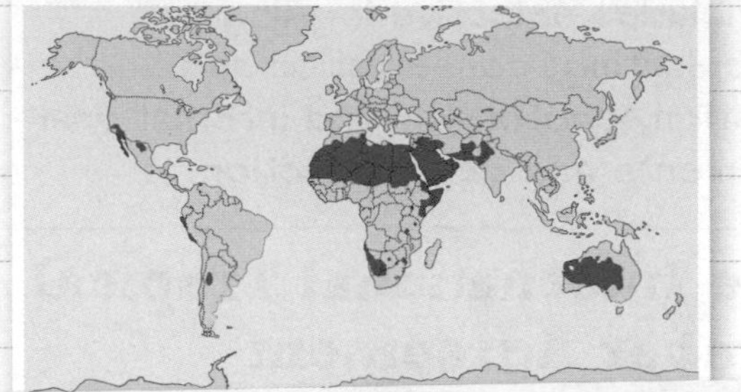

There are different kinds of desert: a hot desert is characterised by high average temperatures and very low precipitation. Hot deserts are located at the subtropics, around 30° north and south of the Equator.

Why are hot deserts hot and dry?

The general atmospheric circulation model explains hot desert characteristics. For more on the general atmospheric circulation model, see page 8.

- Air at the Equator rises 15 km into the atmosphere and travels north and south to around 30° of latitude, where it sinks.
- Hot deserts occur at this zone of dry, sinking air. Sinking air creates high pressure, which gives cloudless conditions.
- Energy from the Sun is not quite as intense as at the Equator, but the clear skies and lack of humidity produce very high daytime temperatures.

Worked example

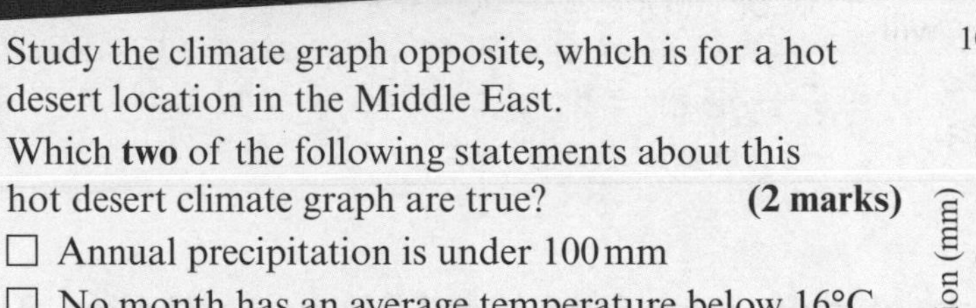

Study the climate graph opposite, which is for a hot desert location in the Middle East.

Which **two** of the following statements about this hot desert climate graph are true? **(2 marks)**

☐ Annual precipitation is under 100 mm

☐ No month has an average temperature below 16°C

☐ No month has an average temperature below 30°C

☐ The lowest precipitation in this Northern Hemisphere location is in the winter months

Multiple choice questions will usually involve interpreting data. Check carefully.

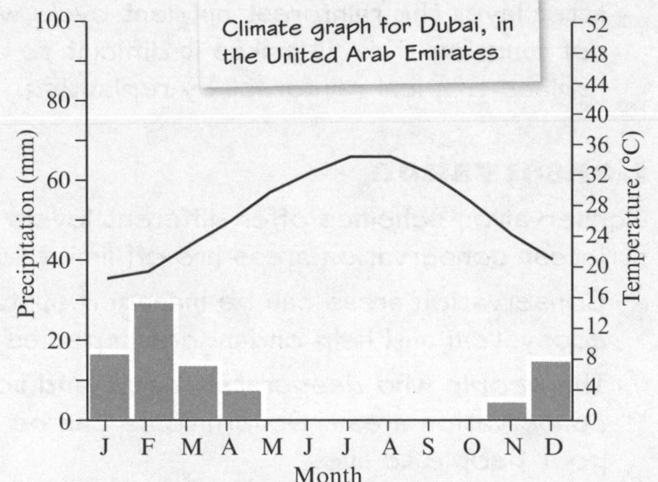

Hot desert characteristics

- ✓ Hot deserts have less than 250 mm of precipitation a year.
- ✓ Daytime temperatures are usually above 20°C all year round, and in the hottest season can often reach 45°C.
- ✓ Most hot desert ecosystems have stony soil, not sand – the sand gets blown away.
- ✓ Plants usually grow close to the ground – there is no canopy.
- ✓ The heat means that rain sometimes evaporates before it reaches the ground.

Now try this

Hot desert ecosystems are located in high-pressure zones. Why does this explain why deserts are cold at night? **(2 marks)**

The hot desert ecosystem

Hot deserts are challenging environments for animals, plants and people. The lack of water makes them low-nutrient ecosystems, with low biodiversity. Animals and plants that survive in the hot desert have specialised adaptations.

Hot desert nutrient cycle

Nutrient cycle diagram of a hot desert ecosystem

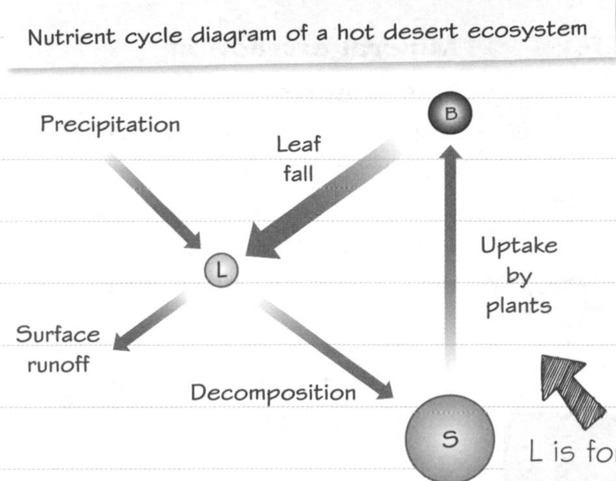

Low-nutrient challenges

- The amount of precipitation coming into the desert ecosystem is low.
- The low amount of rainfall means that decomposition only works very slowly.
- The desert biomass store is small because of the lack of water and low supply of nutrients.
- The heat is also very challenging for plants so only a few species can survive here.
- The soil store is the largest store. This is because there are so few plants taking up nutrients.

L is for **litter store**, S is for **soil store** and B is for **biomass**.

Desert plant adaptations

Desert plants are adapted to hot desert conditions in three main ways.

1. Some species, like the agave, can store water. These are often protected by spines.
2. Some plants have large root networks – either going deep to tap into groundwater (for example, mesquite tree) or wide and shallow (for example, cacti) – to absorb as much water as possible when it does rain.
3. Desert plants, such as poppies, have seeds that only germinate when there is rain – they can lie dormant for many years.

Cacti in the Arizona desert

Human impacts can be very damaging in desert ecosystems. The low-nutrient environment means plants are slow to recover from damage; low biodiversity means damage to one species affects all the others.

Desert animal adaptations

Three ways that animals adapt to the conditions:

1. Nocturnal – by being active at night animals can avoid the high daytime temperatures by going underground
2. Getting water from their food rather than needing to drink – the Dorcas gazelle can do this, for example
3. Big ears radiate heat very efficiently

Worked example

Suggest **two** reasons why biodiversity is low in hot desert ecosystems. **(2 marks)**

There is not enough water or nutrients in the ecosystem to support many plants or animals, which means low biodiversity.

Desert plants and animals have to be highly specialised to survive the high temperatures and lack of water. Other species cannot survive, which reduces biodiversity.

Now try this

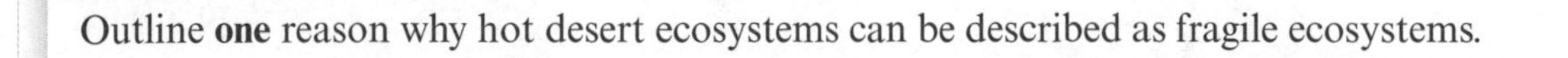
Outline **one** reason why hot desert ecosystems can be described as fragile ecosystems. **(2 marks)**

Development opportunities

You will have done a **case study** of a hot desert to use in answers about development opportunities in hot desert environments: mineral extraction, energy, farming and tourism.

Development opportunities in hot deserts

Farming

The abundant sunshine of hot deserts means that agriculture can be very productive, as long as there is water available for irrigation.

How has farming developed in your case study? What is the source of the water used for irrigation?

Aerial view of development in a desert in the United Arab Emirates

Mineral extraction

Hot deserts often have significant mineral deposits close to the surface because of the way that evaporation pulls water up through the soil, bringing minerals with it.

What are the mineral extraction opportunities in your case study?

Energy

Hot deserts are ideal for generating solar power because of the high intensity sunlight and clear skies. Some countries with large areas of hot desert, such as Saudi Arabia and Iraq, also have large oil reserves.

What are the opportunities for energy production in your case study? Do you know statistics about production?

Tourism

Despite the challenges of living there, hot deserts have many attractions for tourists. They are sunny, have distinctive wildlife and beautiful landscapes, and are exotic. Las Vegas, a town built on entertainment and tourism, is located in the USA's Mojave Desert.

How has tourism developed in your case study of a hot desert?

Worked example

Study the photograph on this page, which shows an aerial view of development in a hot desert in the United Arab Emirates (UAE).

Suggest **one** reason why this area is not being used for agriculture. **(2 marks)**

Although the area would be good for growing crops in terms of heat and light, the main reason is likely to be a lack of water for irrigation.

Suggest means 'present a possible case' or 'present a likely reason why'. You do not have to know the actual answer but instead you use your geographical understanding to make a reasonable, informed suggestion.

Now try this

To what extent does a hot desert environment that you have studied provide opportunities for development? **(9 marks)**

Development challenges

Your **case study** of a hot desert will also have covered challenges of developing hot desert environments: extreme temperatures, water supply and inaccessibility. Use these notes to help with your revision of the case study you did in class.

Extreme temperatures

- Mid-day temperatures in the hot desert can be too hot for people to work, walk, wait for a bus, or go to school.
- The extreme heat of hot desert areas puts enormous stress on solar power cells. They have to be cooled with water, increasing costs.
- **Salinisation**: the extreme heat means that water used for irrigation evaporates, leaving behind mineral salts that are toxic to crops. Land has to be abandoned.

Climate change is making hot desert areas hotter, which increases the pressures on people, livestock and crops.

To feed its rapidly growing population, Egypt has irrigated huge areas of desert to use as farmland. However, **salinisation** affects over 30 per cent of irrigated land in Egypt. Is this also a problem in your case study?

Challenges of water supply

- Most hot desert development depends on water from **shallow aquifers**. Many aquifers have already dried up due to overuse. Only drought-resistant crops can then be grown.
- There are often **deep aquifers** 1 km or more below the desert surface. It is possible to bore down and pump the water up, but it is very expensive and uses a lot of energy.
- Some hot desert areas have **desalination** plants, which produce fresh water from salty water. However, desalination uses a lot of energy and is expensive.
- Conflicts between communities over water have increased, and economic development of desert areas is threatened by conflict and political unrest in hot desert regions.
- Climate change is not helping the challenges of water supply because it is making hot desert areas drier.

Getting enough water is the main challenge of hot desert development. What have been the challenges in your case study?

Inaccessibility

- Mines in hot desert areas are often inaccessible: for example, mines in the Sahara are hundreds of kilometres from Mediterranean ports.
- Workers have to fly to the mines, and their food, water, equipment and fuel have to be transported long distances.
- Deserts are tough places for technology like solar power panels, wind turbines or pipelines. Frequent maintenance is needed, which is difficult in inaccessible locations.

The longest conveyor belt in the world is in Western Sahara. It is 97 km long and carries phosphate ore from deep in the desert to the Atlantic coast.

Now try this

'The biggest challenge to developing hot desert environments is water supply.'
Support this statement with evidence from a hot desert environment you have studied.

(6 marks)

Desertification: causes

Desertification is the process by which land is gradually turned into desert. The main causes of desertification you need to know about are: climate change, population growth, removal of fuelwood, overgrazing, over-cultivation and soil erosion.

Desertification and degraded land

Semi-arid areas, or drylands, are the areas at risk of desertification. It is not usually a process of sand dunes blowing over an area – most desertification happens because different pressures on the land cause it to become **degraded**.

Desertification happens in wealthy countries as well as in poorer countries.

Soil erosion

Soil erosion is the key feature of desertification. Vegetation is very important in drylands for binding soil and returning nutrients to the soil. When vegetation cover is reduced, the topsoil loosens and dries, and is then vulnerable to erosion by wind and water. Without topsoil, the soil that is left is a mixture of dust, sand and stones – no good for plants.

Causes of desertification

Demand for **fuelwood** from increasing population means that trees and bushes are cut down. Without vegetation to protect it, the topsoil is eroded by wind and water.

Increasing populations in dryland regions mean more pressure on soil and water resources. **Population growth** is a major cause of desertification, and also contributes to other causes.

Over-cultivation of dryland regions can use up all the nutrients in the soil. Without soil nutrients, plants cannot grow. Without vegetation, the soil dries out and is eroded.

Increasing numbers of grazing livestock can lead to **overgrazing** – the vegetation binding the soil is eaten and the soil is eroded.

Desert challenges

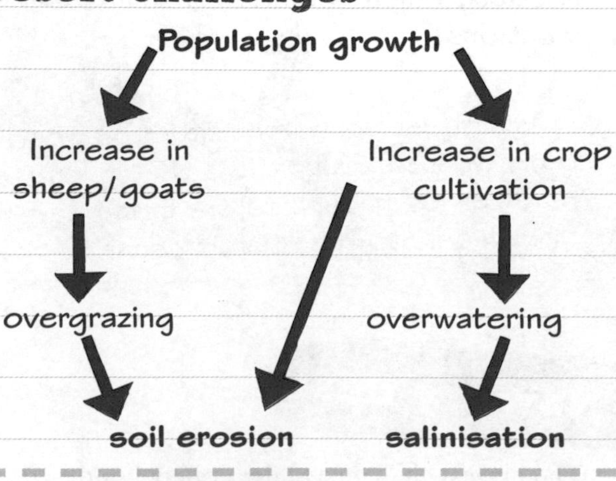

Worked example

Suggest **one** reason why climate change can be a cause of desertification. **(2 marks)**

One consequence of climate change seems to be longer droughts and more unpredictable rainfall in dryland areas. Long droughts mean dryland vegetation dies, removing the protection it gives to the soil. The dry, loose topsoil is then eroded, resulting in desertification.

Now try this

Outline **one** possible impact of population growth in dryland regions of the world. **(2 marks)**

Desertification: reducing the risk

You need to know about strategies used to reduce the risk of desertification. These are: water and soil management, tree planting and use of appropriate technology.

Water management

- Improving water storage: especially during intense showers as this can then help prevent flash flooding and erosion.
- Introducing new ways of harvesting water – for example nets that collect water from fog and dew.
- Salinisation is often reduced by improving the drainage of irrigated fields, so water does not sit on the surface.
- Crops that tolerate less water can replace water-hungry crops like rice or cotton.

Soil management

- Livestock farming and crop farming can be integrated so a farming family uses manure from their animals on their fields to add nutrients to the soil.
- Crop rotations can be introduced so that soil nutrients are not exhausted and so fields are not left bare of vegetation.
- Placing lines of stones along the contours of fields helps prevent erosion, as do terraces – changing sloping fields into a series of flat, stepped areas, which hold on to water better.

Tree planting

Planting trees helps stabilise the soil, increases soil nutrients and protects the soil from wind and water erosion. Planting a mix of tough grasses, bushes and trees can restore degraded land.

As well as **reforestation** – the replanting of deforested land – there is also **afforestation**: planting trees where there were no trees before.

The 'Great Green Wall' project on the southern border of the Sahara is a plan for an 8000 km-long, 15 km-wide 'wall' of trees and bushes aimed at reducing desertification and increasing biodiversity.

Use of appropriate technology

Appropriate technology is suited to the needs of local people and their environments. It is simple enough for local people to be able to use and repair it. For example, well covers to reduce evaporation loss, rainwater harvesting technology and small-scale dams and irrigation methods.

Many of the techniques used to reduce the risk of desertification are adapted from traditional practices of drylands peoples.

Every year, 12 million hectares of land are lost to desertification, affecting 1.5 billion people – many of them among the world's poorest.

Worked example

Study the image of the RainSaucer (right). It has a wide surface area to catch rain and funnel it into a container. Outline **one** reason why this is appropriate technology. **(2 marks)**

The RainSaucer has a simple design. People in dryland areas would be able to fix the RainSaucer if it broke, using locally available materials.

Now try this

One method of reducing the risk of desertification is to grow trees among the crops in dryland fields. Suggest **one** reason why this method helps reduce desertification risks. **(2 marks)**

Cold environments: characteristics

Cold environments include both polar environments and tundra environments. You need to know the main physical characteristics of cold environments.

Only revise cold environments if you studied them in class.

Global distribution of cold environments

Polar cold environment

Tundra cold environment

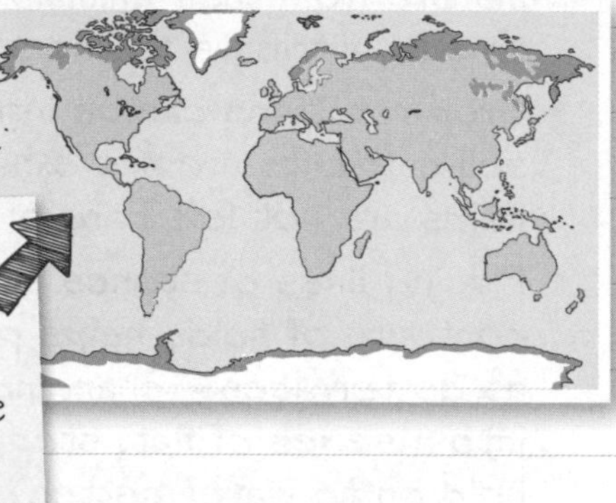

The ice caps are in white; the other cold environments are called tundra

Why so cold?

While the Sun's radiation is intense at the Equator, at the Earth's poles it is less intense because it does not hit at a direct angle.

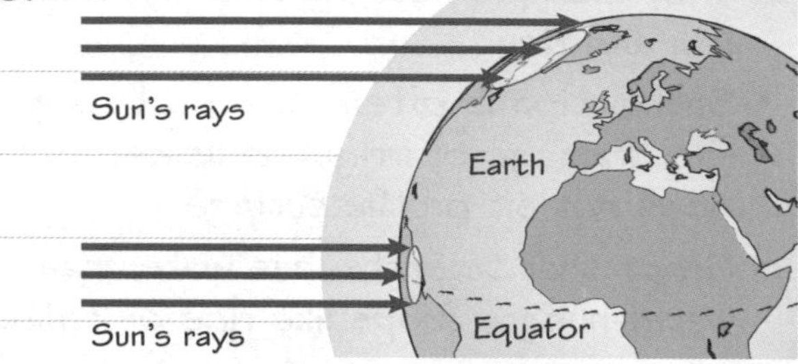

Cold environment climate

Cold environments experience freezing temperatures (below 0°C) for long periods of the year.

There are many different kinds of cold environment. **Extreme cold environments** experience very low winter temperatures (−50°C) and on land they have permanent ice cover. **Polar environments** are extreme cold environments. The **tundra** is a less extreme cold environment, but it is still very cold in winter: −30°C. Summers are cool (around 10°C) and short. Although the top layer of soil unfreezes in summer, underground the **permafrost** remains frozen.

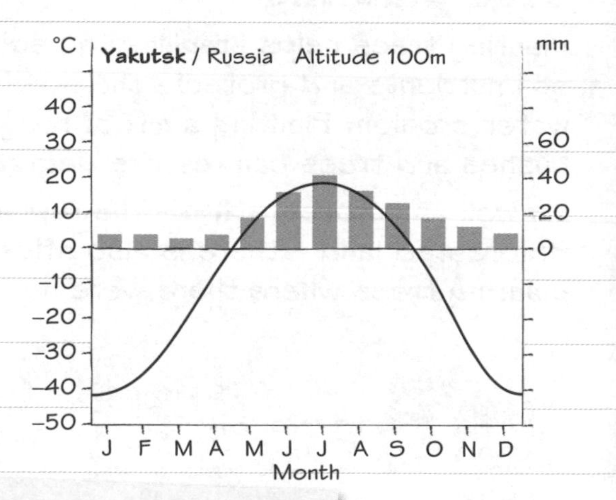

Climate graph for a cold environment

Worked example

Study this photo of a tundra environment. Which **two** of the following statements are correct? **(2 marks)**

☐ **A** It is too cold in the tundra for plants to grow.

☒ **B** Permafrost under the ground means the tundra is boggy.

☐ **C** The tundra cold environment is dominated by coniferous trees.

☒ **D** Plants that can survive the tundra environment include mosses and grasses.

Now try this

Polar and tundra cold environments usually have very low precipitation. Calculate the annual precipitation shown in the climate graph on this page. **(1 mark)**

The cold environment ecosystem

Cold environments are challenging environments for animals, plants and people – sometimes too challenging for all forms of life. Cold, dark and dry conditions mean cold environments are low nutrient ecosystems, with low biodiversity. Animals and plants that can survive all year have specialised adaptations.

A cold environment nutrient cycle

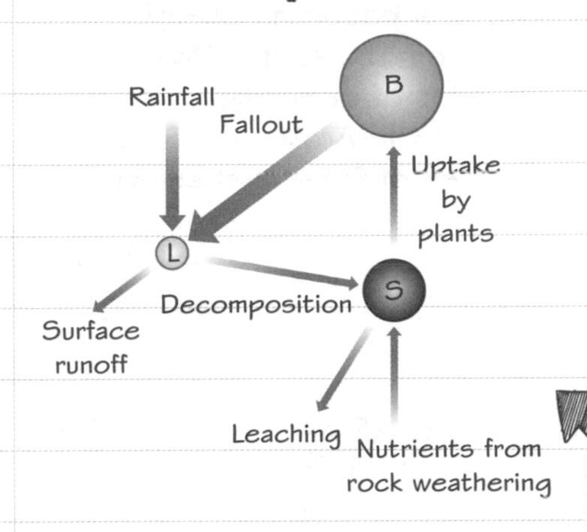

Low nutrient challenges

- The period when it is warm enough and light enough for plants to grow is short: two or three months in summer. This keeps biomass and nutrients at low levels.
- Decomposition cannot happen in freezing conditions and needs water. Decomposition only happens very slowly in the cold, dry conditions of the tundra.
- The litter store is the largest store in cold environments because of the low rate of decomposition. The soil is low in nutrients.
- Plants that do grow, often grow very slowly.

L is for **litter store**, S is for **soil store** and B is for **biomass**.

Cold environment plant adaptations

Tundra plants are adapted to cold environment conditions in three main ways.

1. Plants have shallow roots because only the top layer of the soil unfreezes in summer; permafrost beneath blocks deeper roots.
2. Plants stay dormant through the dark winter but burst into life rapidly as soon as it becomes lighter and warmer. Many plants produce flowers very quickly to maximise pollination time.
3. Plants are low-lying so they are not damaged by the wind. Low nutrient soils also mean plant growth is restricted.

Cold environment animal adaptations

Animals that stay in the cold environment all year have specialist adaptations including:

- good insulation – dense fur and insulating fat layers
- the ability to hibernate/become dormant over winter – lowers metabolic rate to use very little energy
- white fur – provides camouflage against snow and ice and better insulation.

In the short summer months, cold environments are often filled with life as migrating birds travel north. Migrating species then fly south for a warmer climate in the winter.

Worked example

Suggest **two** reasons why biodiversity is low in cold environments. **(2 marks)**

There is not enough water or nutrients in the ecosystem to support much plant growth, which means low biomass and low biodiversity.

Animals and plants are specialised to survive in cold environments. A small number of specialist species means low biodiversity.

Now try this

Outline **one** reason why cold environments can be described as fragile ecosystems. **(2 marks)**

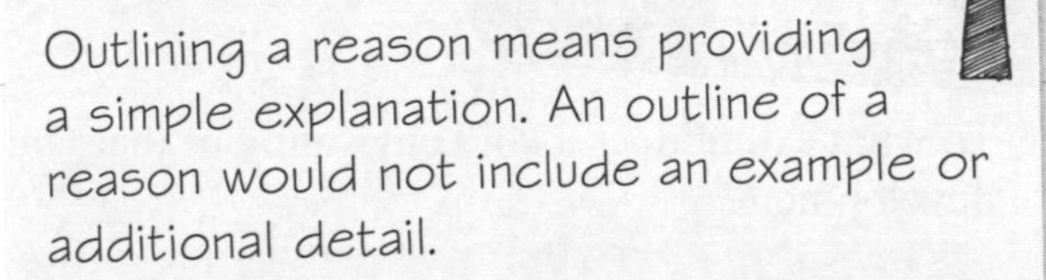

Outlining a reason means providing a simple explanation. An outline of a reason would not include an example or additional detail.

Development opportunities

You will have done a **case study** of development opportunities in a cold environment, which covers: mineral extraction, energy, fishing and tourism.

Development opportunities in cold environments

Mineral extraction

There are large deposits of minerals such as gold, diamonds, uranium and copper in polar regions. The very cold environment makes mining here very expensive. Mining is developed when global prices for the minerals are very high. If prices drop, the development opportunity becomes much less attractive.

What minerals are mined in your case study area? Have global price changes affected mining there?

Energy

The Arctic is thought to have the world's largest undeveloped oil and gas reserves. Now that there is less ice in the Arctic summer due to climate change, it is becoming cheaper to extract oil and gas from the Arctic Ocean. However, there is a lot of opposition to cold environments, such as the Arctic and Antarctic, being used as sources of fossil fuel energy.

What are the opportunities for energy in your case study?

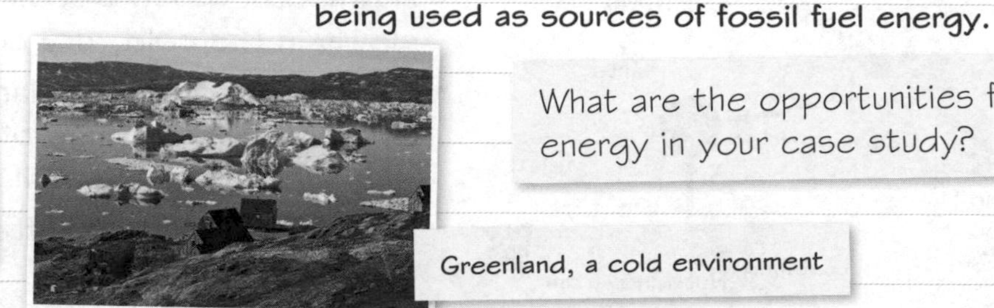

Greenland, a cold environment

Fishing

When ice thaws from the surface of the sea in spring in cold environments, differences in water temperature cause currents to bring nutrient-rich water up from the deep ocean. This triggers an explosion of plankton growth, which attracts huge numbers of fish species. Fishing is important for most coastal communities in cold environments: fish makes up 90 per cent of Greenland's exports.

How important is the fishing industry in your case study cold environment?

Tourism

There are opportunities to develop tourism in cold environments, especially in places where there is distinctive cold environment wildlife. Antarctica is growing in popularity as a tourist destination because of its remoteness and wildness: around 40 000 people a year visit Antarctica. Because of the difficulty of getting people there, it is a very expensive holiday: around £4000 per person.

How important is tourism in your case study area?

Worked example

Study the photograph on this page, which shows a settlement in Greenland. Suggest one reason why this area could be developed for tourism. **(2 marks)**

Tourists would be interested to see the dramatic landscape and the icebergs. This could be a good location for adventure tourism.

Suggest means 'present a possible case' or 'present a likely reason why'. You do not have to know the actual answer but instead you use your geographical understanding to make a reasonable, informed suggestion.

Now try this

To what extent does a cold environment that you have studied provide opportunities for economic development? **(9 marks)**

Development challenges

Your **case study** of a cold environment will also have covered challenges of developing cold environments: extreme temperature, inaccessibility, provision of buildings and infrastructure.

Extreme temperature

Any cold environment can be dangerous if people are not prepared for low temperatures.

- Hypothermia is when body temperature drops too low for normal brain function. People fall asleep and freeze to death.
- Frostbite is another hazard – when body extremities (such as fingers, noses) freeze, damaging the body tissues.
- The lubricants used in machinery behave differently in cold temperatures, becoming sticky and damaging machinery.

Inaccessibility

Remote locations intensify the challenges for development: extreme cold, long periods with no sunlight, frequent storms and having to work on uncertain terrain. Inaccessibility means that:

- remote locations are expensive to reach
- if accidents occur or equipment breaks down, it can be difficult to access health care or spare parts
- developers must take all the food, energy and equipment they will use with them, adding to costs and increasing challenges.

Buildings

Buildings in cold environments have to be specially adapted to the conditions with triple-glazed windows to keep out the cold, well-insulated walls and roofs. Roofs need to have a steep angle to shed heavy snow.

Heating and lighting for houses through the long winters of tundra and polar regions is extremely expensive. This is one reason why development in cold environments is often highly seasonal – it is restricted to the short summer months.

Snow and ice adds to the challenges of inaccessibility; this snowcat ambulance can reach locations that are not accessible to other vehicles

Worked example

Buildings in the Canadian Arctic on stilts, with above-ground water and sewerage pipes covered by conduits

Study the image on the left, which shows buildings and infrastructure in the Canadian Arctic cold environment.

Explain **one** reason why the water and sewerage infrastructure is above ground in this cold environment location. **(2 marks)**

Permafrost causes major problems for infrastructure. Melting and thawing of the top layer of soil causes underground pipes to break, and makes roads buckle and crack. That is why these pipes are above ground.

Now try this

'The biggest challenge to developing cold environments is extreme temperature.'
Support this statement with evidence from a cold environment that you have studied. **(6 marks)**

Fragile wilderness

So many natural environments have been changed by human activity that it is rare for most people to experience wilderness – places still largely unaltered by humans. However, cold wilderness environments are fragile and under threat.

Why are cold environments fragile?

Fragile ecosystems are easy to damage and take a long time to recover.

- Plants grow very slowly because of the lack of nutrients and the cold, dark winters.
- The low rate of nutrient cycling, low precipitation and poor drainage of many cold environments means that pollution remains in the ecosystem for decades.
- There are very few species in tundra and polar environments. A disease that affects one species has an impact on the whole ecosystem.
- Animals and plants are highly specialised. This makes them very vulnerable to change – especially climate change.

Worked example

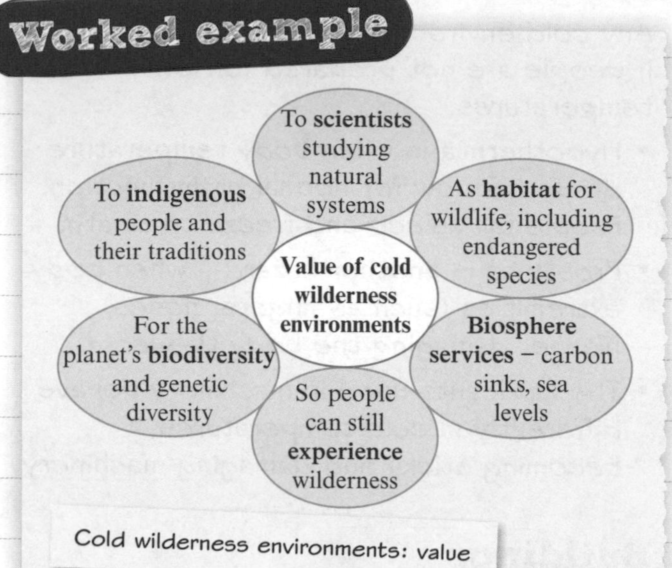

Cold wilderness environments: value

Give **two** reasons why cold environments are valuable as wilderness areas. **(2 marks)**

Wilderness areas have not been altered by human activity. Cold wilderness areas are valuable stores of carbon, locked away in deep layers of peat and so not contributing to climate change. They are also valuable to indigenous peoples, who have lifestyles closely adapted to sustainable use of wild animals.

Threats to cold environments

- **Extracting fossil fuels** – oil spills can poison the fragile cold environment ecosystem for many years.
- **Acid precipitation** – chemicals from industrial pollution combine with rainwater to produce acid rain, which kills cold environment plants and acidifies the lakes and marshes where insects live.
- **Climate change** – warmer temperatures are already disrupting polar habitats, increasing the risk of forest fires and allowing new competitor species, pests and diseases into cold environment ecosystems.

The threat of climate change to cold environments is a global threat, due to the impact of warmer temperatures on ice on land. Melting ice sheets and glaciers will raise sea levels; melting permafrost will release huge amounts of stored carbon into the atmosphere.

Now try this

Study the photo, which shows a tundra landscape from Canada in an area called The Barrenlands. 'Barren' means unproductive and having no value to humans.

Suggest **two** ways in which this wilderness area would now be considered valuable to humans. **(2 marks)**

Use the diagram on this page to help you come up with ideas to answer this question.

Managing cold environments

You need to know about strategies to balance the needs of economic development and conservation in cold environments: use of technology, role of governments, international agreements and conservation groups.

Use of technology

- **Green roofs** contain a deep layer of soil with hardy cold environment plants – they provide good heat insulation while helping reduce the impact of buildings in wilderness areas.
- Salt is used to melt ice on roads, but it dehydrates plants and poisons drinking water for animals. **'Smart' snowploughs** use sensors to calculate the minimum effective amount of salt needed to clear the ice.
- **Thermosiphons** are heat-exchange devices. Stuck into the ground, they pull out heat from permafrost to keep it frozen for longer, reducing damage to infrastructure and buildings from melting soil movement.

A thermosiphon in Alaska, which reduces 'frost heave' damage

Governments

- Governments can make areas of cold environment into national parks or protected wilderness reserves.
- Some reserves ban economic development and focus on **conservation** and scientific research.
- Other reserves strictly control the type of economic development allowed and may insist on **sustainable resource** use.
- Governments can pass laws to protect the **rights of indigenous peoples**. For example, the Constitution Act of 1982 recognises the right of Canada's indigenous peoples to land.

Conservation groups

Conservation groups monitor the impacts of development and pressure governments to protect vulnerable ecosystems. These groups:

- **campaign** to end drilling for oil in sensitive polar environments, such as Lancaster Sound in the Canadian Arctic
- **raise money** to support sustainable developments for local communities (for example, ecotourism)
- **provide funds and expertise** for restoration and rapid responses to environmental emergencies, such as oil spills.

Worked example

Read the Fact file about the Antarctic Treaty. Suggest **two** ways in which this treaty has reduced the impact on the polar environment of the developing Antarctic tourism industry.

(2 marks)

Ships have to use less polluting fuels (not heavy diesel) to reduce the impact of marine pollution. Tourists cannot leave any litter when they land in case this affects local ecosystems.

Fact file: The Antarctic Treaty

The Antarctic Treaty protects the polar environment of the Antarctic continent. It:

- bans all mineral extraction
- requires all activities in the Antarctic to complete an Environmental Impact Assessment before they are permitted
- sets out strict procedures for waste disposal so that local habitats are not disturbed
- strictly controls pollution from boats.

Now try this

Suggest **one** reason why tourism operators are not allowed to land more than 100 tourists on Antarctica at a time.

(2 marks)

Had a look ☐ Nearly there ☐ Nailed it! ☐

Physical landscapes in the UK

In this section, you need to revise this page (the major upland and lowland areas of the UK and the main river systems) and **two** from UK coastal landscapes, UK river landscapes and UK glacial landscapes. See page 45 for UK coastal landscapes, page 54 for UK river landscapes and page 64 for UK glacial landscapes.

UK upland areas

The main upland areas in the UK are in the north and west.

UK lowland areas

The main lowland areas of the UK are in the east and the south.

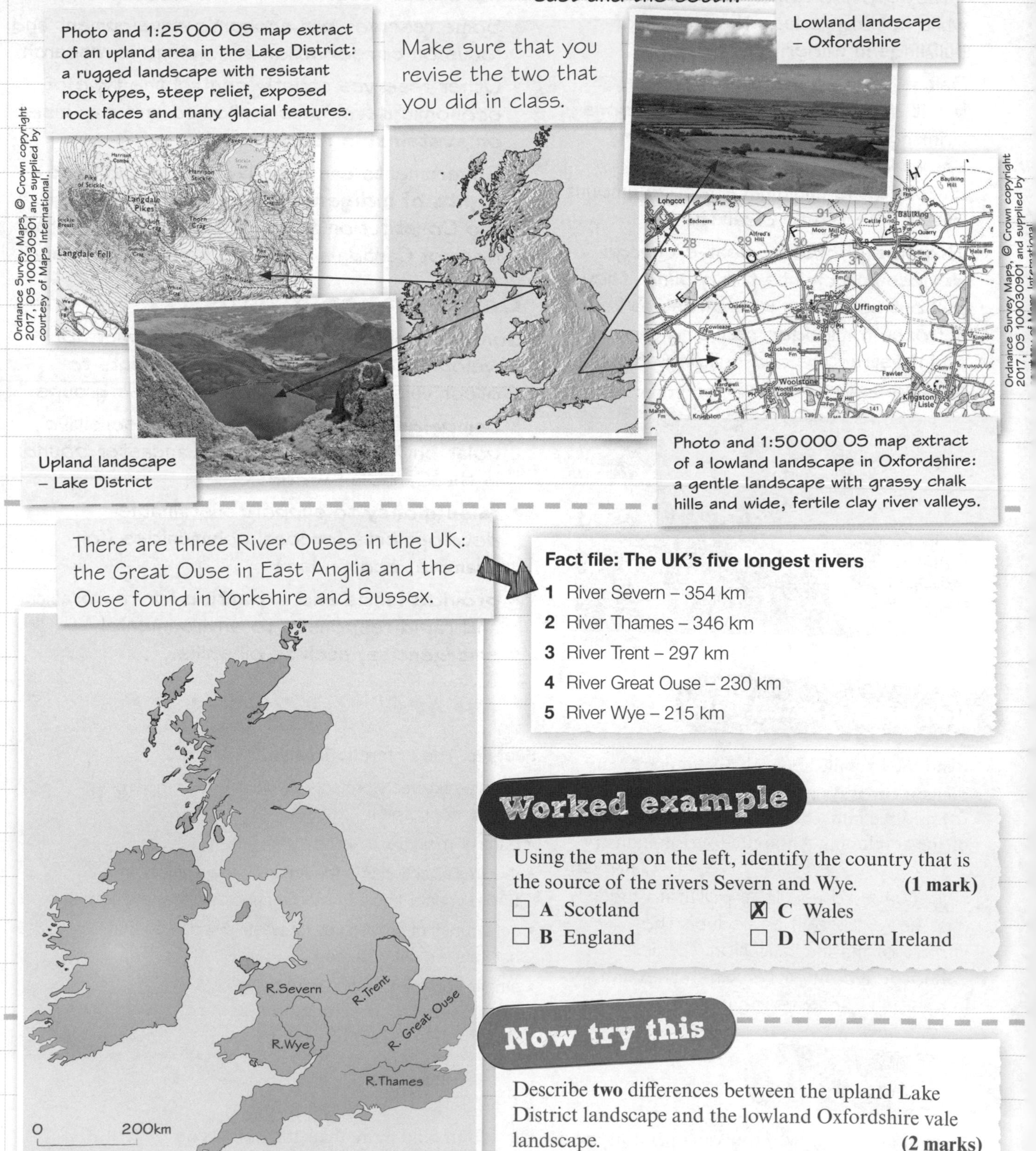

Photo and 1:25 000 OS map extract of an upland area in the Lake District: a rugged landscape with resistant rock types, steep relief, exposed rock faces and many glacial features.

Make sure that you revise the two that you did in class.

Lowland landscape – Oxfordshire

Upland landscape – Lake District

Photo and 1:50 000 OS map extract of a lowland landscape in Oxfordshire: a gentle landscape with grassy chalk hills and wide, fertile clay river valleys.

There are three River Ouses in the UK: the Great Ouse in East Anglia and the Ouse found in Yorkshire and Sussex.

Fact file: The UK's five longest rivers

1. River Severn – 354 km
2. River Thames – 346 km
3. River Trent – 297 km
4. River Great Ouse – 230 km
5. River Wye – 215 km

Worked example

Using the map on the left, identify the country that is the source of the rivers Severn and Wye. **(1 mark)**

☐ **A** Scotland
☐ **B** England
☒ **C** Wales
☐ **D** Northern Ireland

Now try this

Describe **two** differences between the upland Lake District landscape and the lowland Oxfordshire vale landscape. **(2 marks)**

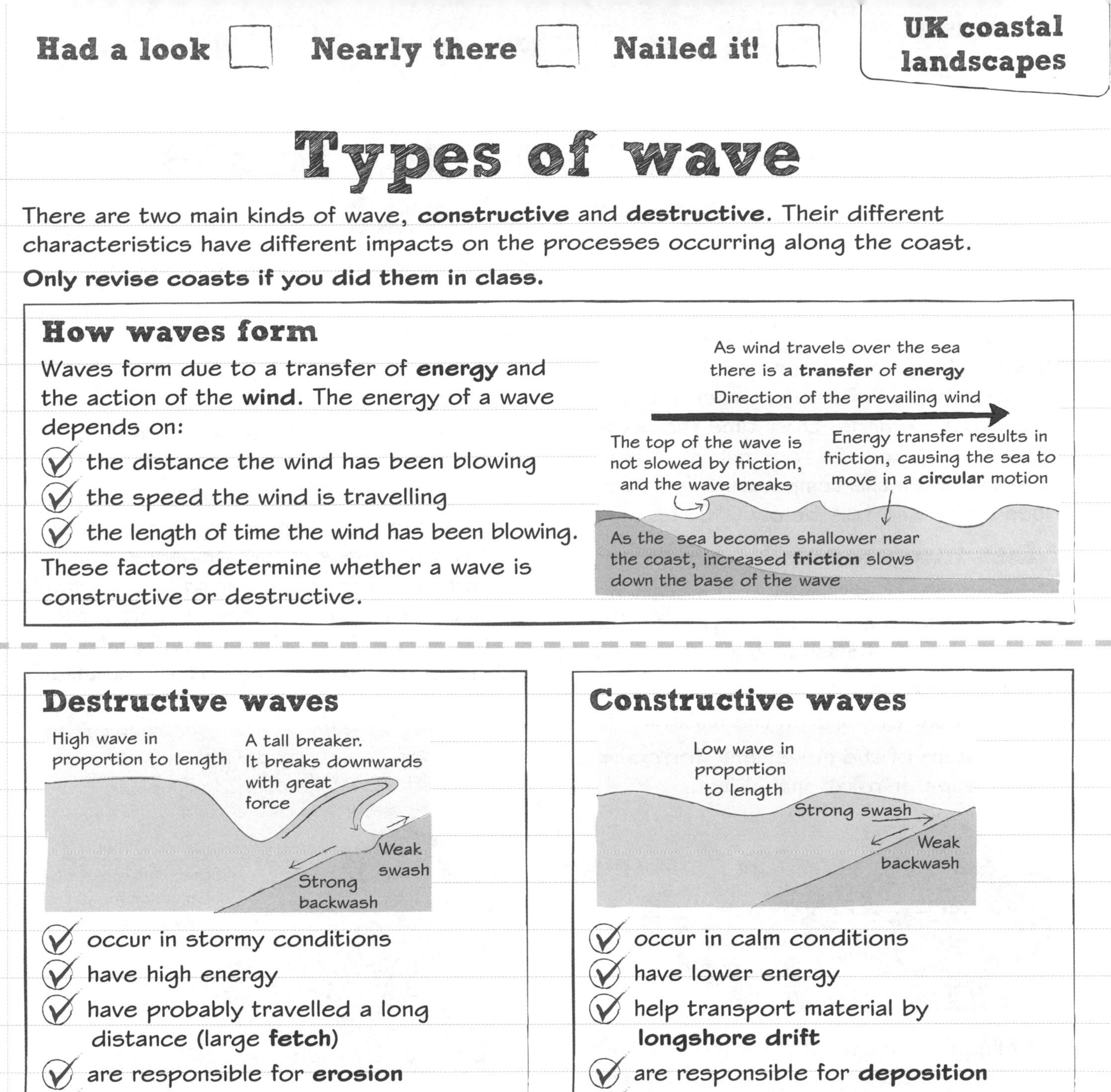

Types of wave

There are two main kinds of wave, **constructive** and **destructive**. Their different characteristics have different impacts on the processes occurring along the coast.

Only revise coasts if you did them in class.

How waves form

Waves form due to a transfer of **energy** and the action of the **wind**. The energy of a wave depends on:

- ✓ the distance the wind has been blowing
- ✓ the speed the wind is travelling
- ✓ the length of time the wind has been blowing.

These factors determine whether a wave is constructive or destructive.

Destructive waves

- ✓ occur in stormy conditions
- ✓ have high energy
- ✓ have probably travelled a long distance (large **fetch**)
- ✓ are responsible for **erosion**
- ✓ have a greater **backwash** than **swash**.

Constructive waves

- ✓ occur in calm conditions
- ✓ have lower energy
- ✓ help transport material by **longshore drift**
- ✓ are responsible for **deposition**
- ✓ have a greater **swash** than **backwash**.

Worked example

Complete this statement about the diagram opposite.

(2 marks)

Apart from being less powerful than destructive waves, another feature of <u>constructive</u> waves is that the stronger <u>swash</u> deposits material up the beach rather than eroding the beach.

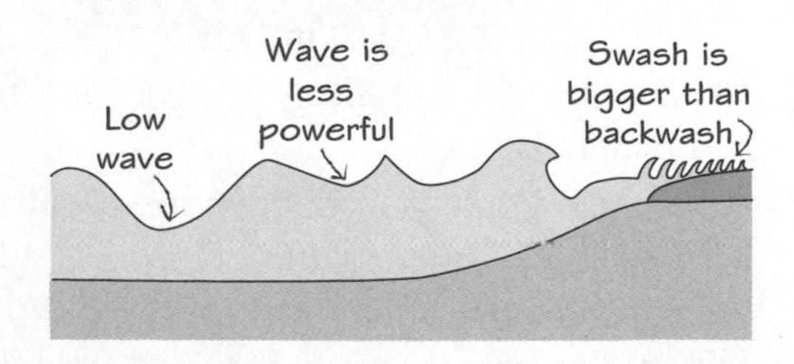

Now try this

Outline the characteristic features of a destructive wave. You may use a diagram in your answer. **(4 marks)**

If you use a diagram as part of your answer, it should be **annotated** to show the features.

Weathering and mass movement

Processes that impact on the land – such as weathering and mass movement – also contribute to coastal erosion.

Mechanical weathering

Freeze–thaw – most common in **cold** climates. When it freezes, water in cracks in the rock expands. Over time the crack widens and pieces of rock fall off. It is most effective when the temperature frequently rises above and falls below 0°C.

Mass movement

Mass movement is the downhill movement of material under the influence of gravity. The different types of mass movement depend on:

- the material involved
- the amount of water in the material
- the nature of the movement, for example, falls, slips or rotational slides.

Chemical weathering

This happens when the rock's mineral composition is changed.

- Granite contains feldspar which converts to soft clay minerals as a result of a chemical reaction with water.
- Limestone is dissolved by **carbonation.** Carbon dioxide in the atmosphere combines with rainwater to form carbonic acid, which changes calcium carbonate (limestone) into calcium bicarbonate. This is carried away by water in **solution.**

Slumps happen when the rock (often clay) is saturated with water and slides down a curved slip plane.

Direction of slide
Slide plane

Sliding happens when loosened rocks and soil suddenly tumble down the slope. Blocks of material might all slide together.

Loose, wet rocks slump down under the pull of gravity along curved slip planes

Worked example

Study the photo, which shows a rock fall on chalk cliffs along the Jurassic Coast in Dorset, UK.
Explain the processes likely to have contributed to this rock fall. **(4 marks)**

Rock falls usually occur where a cliff face has lots of joints and cracks. Repeated freeze–thaw weathering gradually loosens blocks of rock, which fall to the cliff base. If the sea erodes the base of the cliff this can also make rock falls more likely.

A rock fall on the south-west coast near Old Harry Rocks, Jurassic Coast, Dorset, UK

Now try this

Study the photo in the Worked example. Apart from freeze–thaw weathering, state **one** other form of weathering that could affect these cliffs. **(1 mark)**

Erosion, transport, deposition

Waves are responsible for eroding and transporting material. When waves lose the energy that they need to continue carrying material, they deposit the material.

Processes of coastal erosion

- **Hydraulic power**: the weight and impact of water against the coastline, particularly during a storm, erodes the coast. Waves also compress air in cracks in the rock, forcing them apart and weakening the rock.
- **Abrasion**: during storms, breaking waves throw sand, pebbles or boulders against the coast.
- **Attrition**: rocks and pebbles carried by waves rub together and break down into smaller pieces.
- **Solution**: seawater causes some rocks, especially limestone, to gradually dissolve.

Destructive waves and erosion

Destructive waves are the main eroding force along the coastline. In a destructive wave, the swash is weak and the backwash is strong, which means material is dragged back down a beach and into the sea, instead of being pushed onto it.

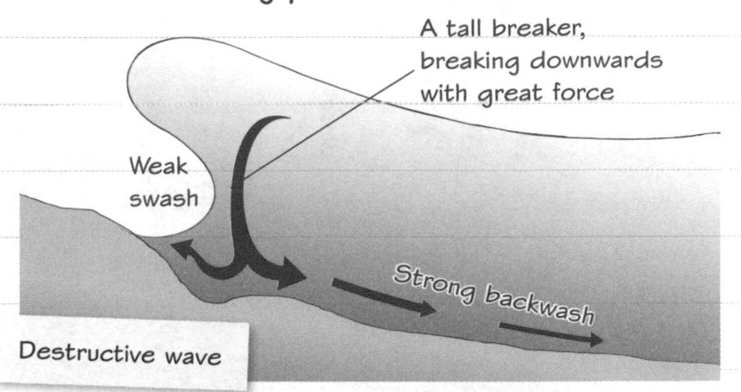

Destructive wave

Longshore drift

1. Waves approach the coast at an angle.
2. **Swash** pushes sand and gravel up the beach at the same angle.
3. **Backwash** carries sand and gravel back down the beach at 90° to the coastline under the force of gravity.
4. Sand and gravel move along the beach in a zigzag fashion.
5. Sand is lighter than gravel so moves further up the beach.

Sediment
Backwash
Beach
Swash
Longshore drift
Wave front
Sea
Prevailing wind direction

Adapted diagram courtesy of Barcelona Field Studies Centre, www.geographyfieldwork.com

Deposition

The load carried by waves is deposited by constructive waves. Different factors influence deposition, such as:

- sheltered spots (for example, bays)
- calm conditions
- gentle gradient offshore, causing friction.

All reduce the wave's energy.

Worked example

What is the main way in which deposition occurs? **(1 mark)**

☐ **A** Waves have high energy, causing material to be dropped.

☒ **B** Waves have lower energy, causing material to be dropped.

☐ **C** Waves remove material from the coastline.

☐ **D** Wind is mainly blowing offshore.

Now try this

What is the term used to describe the direction from which the wind most often blows at a coastline? **(1 mark)**

Erosion landforms

The type of rock at the coast and the structure of the coast's geology both influence the coastal landforms produced by erosion: headlands and bays, cliffs and wave-cut platforms, caves, arches and stacks.

The formation of headlands and bays

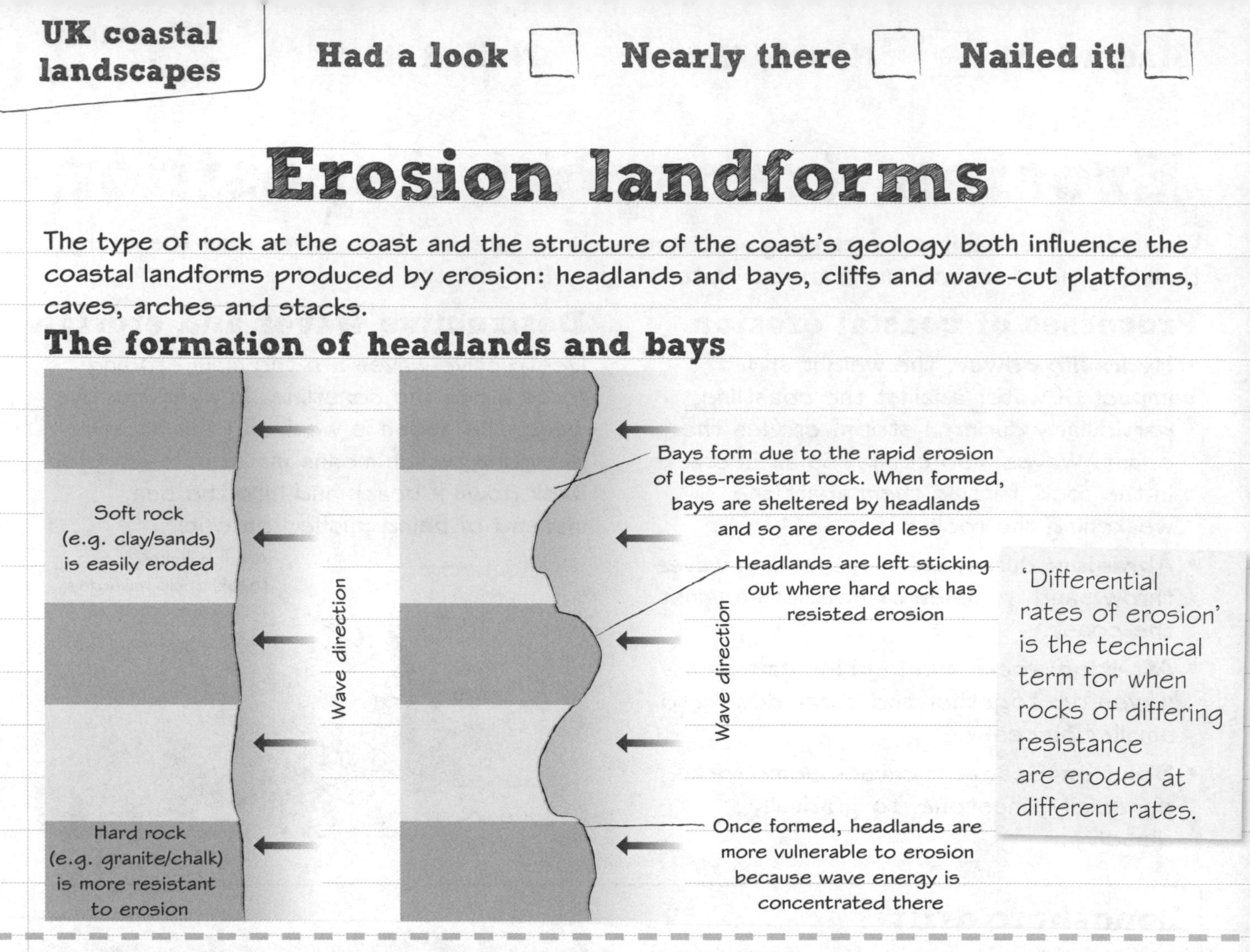

'Differential rates of erosion' is the technical term for when rocks of differing resistance are eroded at different rates.

Hard rock coastal landforms created by erosion

Caves, arches and stacks

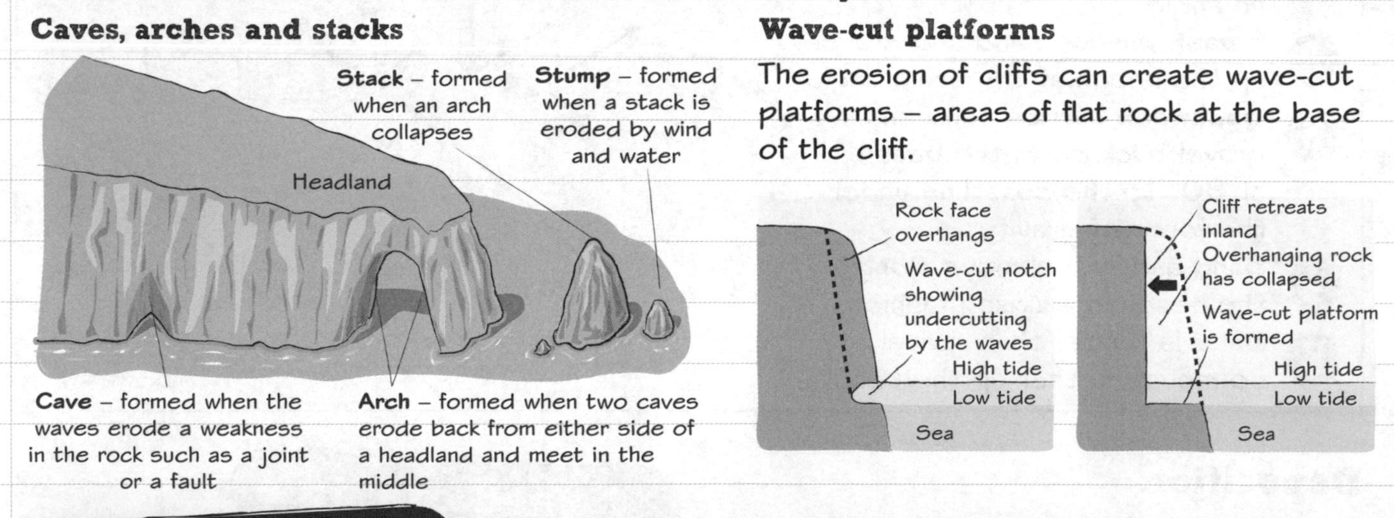

Wave-cut platforms

The erosion of cliffs can create wave-cut platforms – areas of flat rock at the base of the cliff.

Worked example

Explain how geological structure can influence the erosion of a coastal headland. **(4 marks)**

Headlands are sections of resistant rock which jut out in the sea. Erosion by the sea will happen faster where there are gaps, cracks, joints, faults or other weaknesses in the rock. Caves are formed as erosion happens more quickly at this weaker section of the rock. If the weakness goes right through the rock, an arch may form as caves on either side of the headland join up.

Now try this

You could use diagrams to help you answer questions such as this one.

Explain how a wave-cut platform is formed. **(4 marks)**

Deposition landforms

Landforms that result from deposition include beaches, sand dunes, spits and bars.

Beaches

Beaches are accumulations of sand and shingle formed by deposition and shaped by erosion, transportation and deposition.

Beaches can be straight or curved. Curved beaches are formed by waves refracting, or bending, as they enter a bay.

Beaches can be sandy or pebbly (shingle). Shingle beaches are usually found where cliffs are being eroded and where waves are powerful. Ridges in a beach parallel to the sea are called berms and the one highest up the beach shows where the highest tide reaches.

Material moved along beach in a zigzag way by longshore drift

Coastline changes direction

Spit curves with change of wind direction

Prevailing winds bring waves in at an angle

Spit

Spits are narrow projections of sand or shingle that are attached to the land at one end. They extend across a bay or estuary or where the coastline changes direction. They are formed by longshore drift powered by a strong prevailing wind.

Bars form in the same way as spits, with longshore drift depositing material away from the coast until a long ridge is built up. However, bars grow right across the bay, cutting off the water to form a lagoon.

Worked example

The diagram on the right shows sand dune succession. Using this diagram, explain the processes involved in the formation of sand dunes. **(4 marks)**

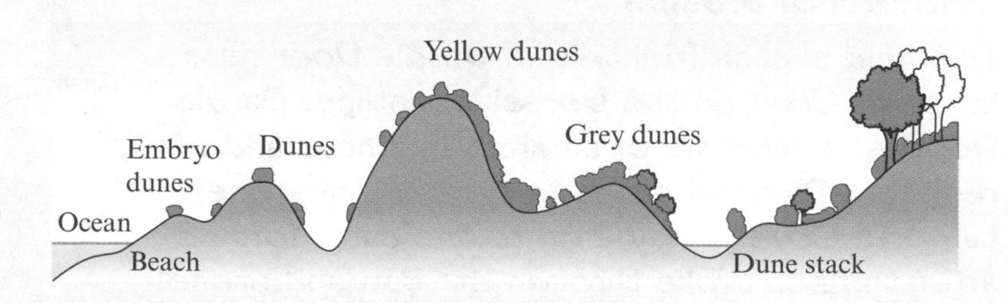

Sand brought onto the beach by longshore drift is blown up the beach by the wind. Where sand gets trapped against something, like a stone, an embryo dune may form. These grow bigger as more sand is deposited. Vegetation starts to grow on the dunes and stabilises them. The mature dunes, furthest up the beach, start to develop soils and support different plant species.

Now try this

Explain why obstacles cause sand to be deposited by the wind to form sand dunes. **(4 marks)**

Coastal landforms

You will have studied an **example** of a section of UK coastline and identified landforms of erosion and deposition along it. **You should revise the example that you studied in class.**

Worked example

Using the diagram on the right and your own knowledge, explain how rock type and geological structure affect the formation of headlands and bays. **(4 marks)**

Headlands and bays form on coastlines where there is a mixture of hard (resistant) and soft rock. Resistant rocks such as limestone or chalk take a long time to erode and are often left jutting out into the sea as headlands. An example is Ballard Point, near Swanage. Soft rocks such as gravels, sands and clay are less resistant to erosion and wear away more quickly to form bays, such as Swanage Bay.

The Isle of Purbeck's discordant coastline (Dorset), where bands of hard and softer rock produce headlands and bays

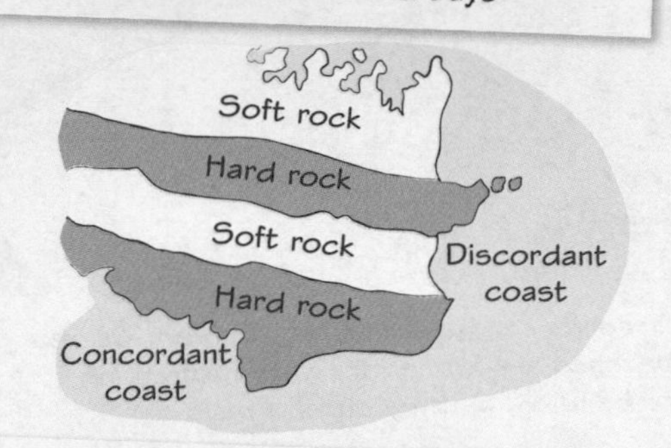

How do rock type and geological structure affect your coastline example?

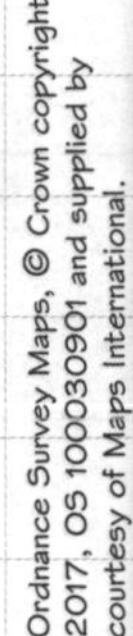

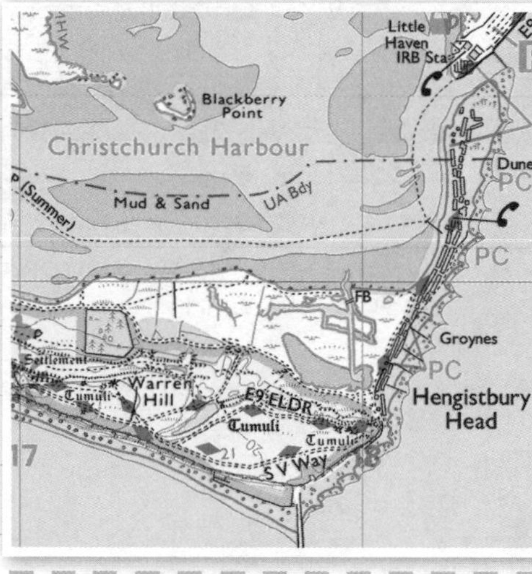

Example: Landform of deposition

An example from the Dorset coastline is Mudeford Spit. This spit is made up of sand and shingle from Hengistbury Head being transported eastwards along the coastline by longshore drift. The spit has formed where the coastline changes direction.

What is your example of a deposition landform? What details do you know about it?

Landform of erosion

This field sketch (right) is of Durdle Door near Lulworth Cove on the Dorset coastline. Durdle Door is an example of an arch, in a headland of resistant Portland stone (a type of limestone). Lulworth Cove, a bay 2 km to the east, formed after the sea broke through the Portland stone and eroded into the soft clays behind it.

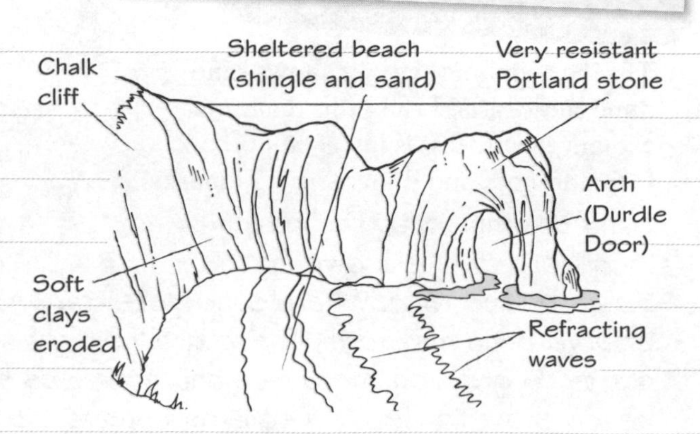

What is your example of an erosion landform? What details do you know about it? Draw a diagram of it.

Now try this

Study the OS map extract and aerial photo of Mudeford Spit. Identify the hard-engineering structures that have been used on the spit to try to reduce its erosion. **(1 mark)**

Hard engineering

You need to know the costs and benefits of the following hard-engineering strategies: sea walls, rock armour, gabions and groynes.

Sea wall

Sea walls are made of concrete or rock; the curved face on some sea walls is designed to reflect wave energy

- £5000–£10 000 per metre
- Reliable protection for buildings and cliffs
- Very expensive to construct and to maintain

Rock armour

Rock armour uses big blocks of highly resistant rock to absorb wave energy

- £1000–£3000 per metre
- Quick to build and easy to maintain
- Rock armour often looks messy; blocks often need to be repositioned

Gabions

Gabions are wire baskets filled with stones; they are only used on sandy beaches: waves carrying shingle would quickly break them

- £100–£500 per metre
- Cheap and easy to construct – any stones can be used to fill the cages
- The cages can rust and break; broken cages have sharp, rusty, dangerous edges

Groynes

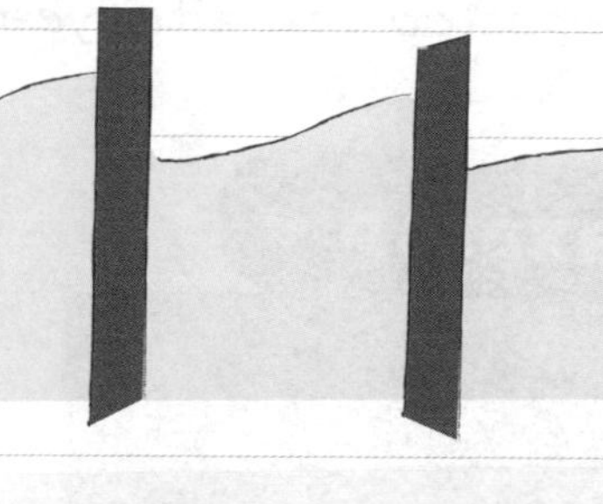

These barriers built out into the sea trap sediment being moved along the beach by longshore drift

- £5000 each (spaced at 200 m)
- Widen the beach: good for tourism
- Longshore drift moves sediment all along the coastline, so groynes on one beach prevent sediment reaching others

Worked example

Study the diagram of a sea wall above. As well as the sea wall, what other hard-engineering strategy is shown in this diagram? **(1 mark)**

A rock groyne

Now try this

Name the erosion process that would be involved if waves carrying shingle broke through wire-mesh gabions. **(1 mark)**

Soft engineering and managed retreat

You need to know the costs and benefits of soft-engineering strategies: beach nourishment and reprofiling, dune regeneration and coastal realignment – a managed retreat strategy.

Beach nourishment

Material is dredged from the sea or brought from other beaches to an eroded beach

- £ Varies according to method used. It cost £200 000 to restore Eastbourne beach after winter storms in 2011.
- 👍 Widens the beach: good for tourism
- 👎 Has to be repeated year after year

Beach reprofiling

Reprofiling at Hythe in Kent; material from the lower part of the beach is moved to the upper part, where it provides better protection

- £ Varies according to project; reprofiling at Hythe costs £250 000 every 2 years
- 👍 Improves protection for seafront property
- 👎 Can be difficult for tourists to get down a steep upper beach for recreation

Worked example

This photo shows coastal realignment at Tollesbury, Essex. Using this photo and your own knowledge, explain the costs and benefits of coastal realignment. **(4 marks)**

Coastal realignment removes coastal protection from areas. It does not cost much, but there are economic costs from loss of earnings from farmland or loss of homes. There are social costs because it is often unpopular with local people. Benefits include the money saved on coastal protection, which can be spent protecting more valuable sites, and the creation of valuable natural habitat in the form of salt marsh.

Dune regeneration

Dune regeneration can involve creating new dunes: for example, by planting marram grass; other dune regeneration methods involve stabilising existing dunes with fencing or adding new sand to dunes that have become eroded

- £ Up to £2000 per 100 m
- 👍 A natural barrier to wind and waves
- 👎 Dunes take a long time to stabilise; regenerated dunes can move or erode again

Now try this

Although coastal realignment is cheap, requires little maintenance and creates extremely valuable habitat for wading birds, it is quite rarely used in the UK. Suggest **one** reason why this might be. **(2 marks)**

Coastal management

You need to know an **example** of a coastal management scheme in the UK so you can write about: the reasons for management, the management strategy and the resulting effects and conflicts. **Make sure you revise the example you learned in class.**

The Holderness coast, east Yorkshire

The mean rate of erosion along this 60 km of coastline is the highest in Europe: 1.8 m per year. The rapid erosion of the boulder clay coast has resulted in the loss of many settlements and farmland over the centuries.

Hard-engineering strategies have been used along the coast. Two rock groynes and rock armour were constructed at Mappleton, costing £2 million.

This management strategy has had a number of results.

- The village of Mappleton (100 inhabitants) is now protected, as is the coastal B1242 road. However, the rock groynes have reduced the supply of sediment south along the coast.
- The small village of Aldbrough, south of Mappleton, has a 'do nothing' management strategy. The beach has narrowed and no longer protects the cliff base.
- The council did not want to build a rock groyne at Hornsea because reducing sediment flow into the Humber could mean erosion of river banks near Hull, threatening the flooding of a major city. Protests from Hornsea residents meant the groyne was built.
- At Easington, south of Mappleton, a terminal handling a quarter of all the UK's gas supply had to be protected with rock armour, costing £4.5 million.

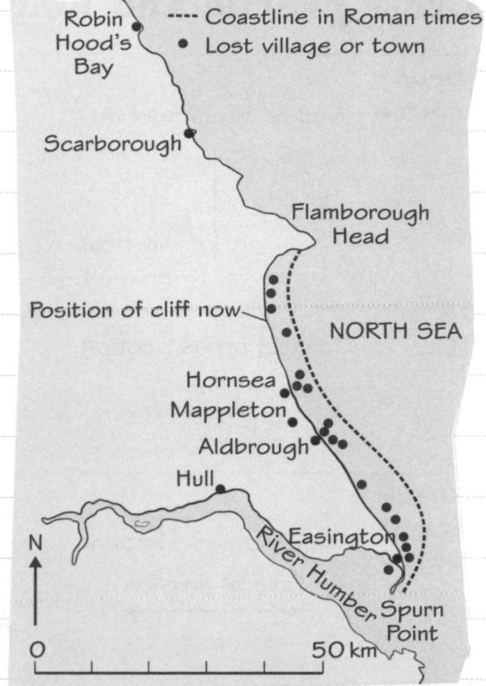

The Holderness coast

Worked example

Explain why there are conflicting views on how a coastline should be managed, using an example that you have studied. **(4 marks)**

The Holderness coast in east Yorkshire is experiencing the fastest cliff retreat in Europe (1.8 m per year on average). There are many conflicting views about how to manage it.

The council is committed to protecting an important gas terminal at Easington, but much of the coastline is under a 'do nothing' coastal strategy because the costs of protecting farmland and property are greater than the benefits of providing protection.

Residents at places like Aldbrough village, without protection, disagree. Although they say that defences at Mappleton have increased erosion at Aldbrough, they still want hard-engineering protection for their community, as the rapid erosion has reduced local house prices. Other council residents, away from the coastline, object to paying more tax to contribute to expensive coastal protection measures that benefit only a few families.

Now try this

Use your own example to write an answer to the Worked example question. **(4 marks)**

River valleys

Rivers, and the valleys they flow in, change in different ways between their source (where they start) and their mouth (where they join the sea). **Only revise rivers if you studied them in class.**

Cross profile and long profile of a river

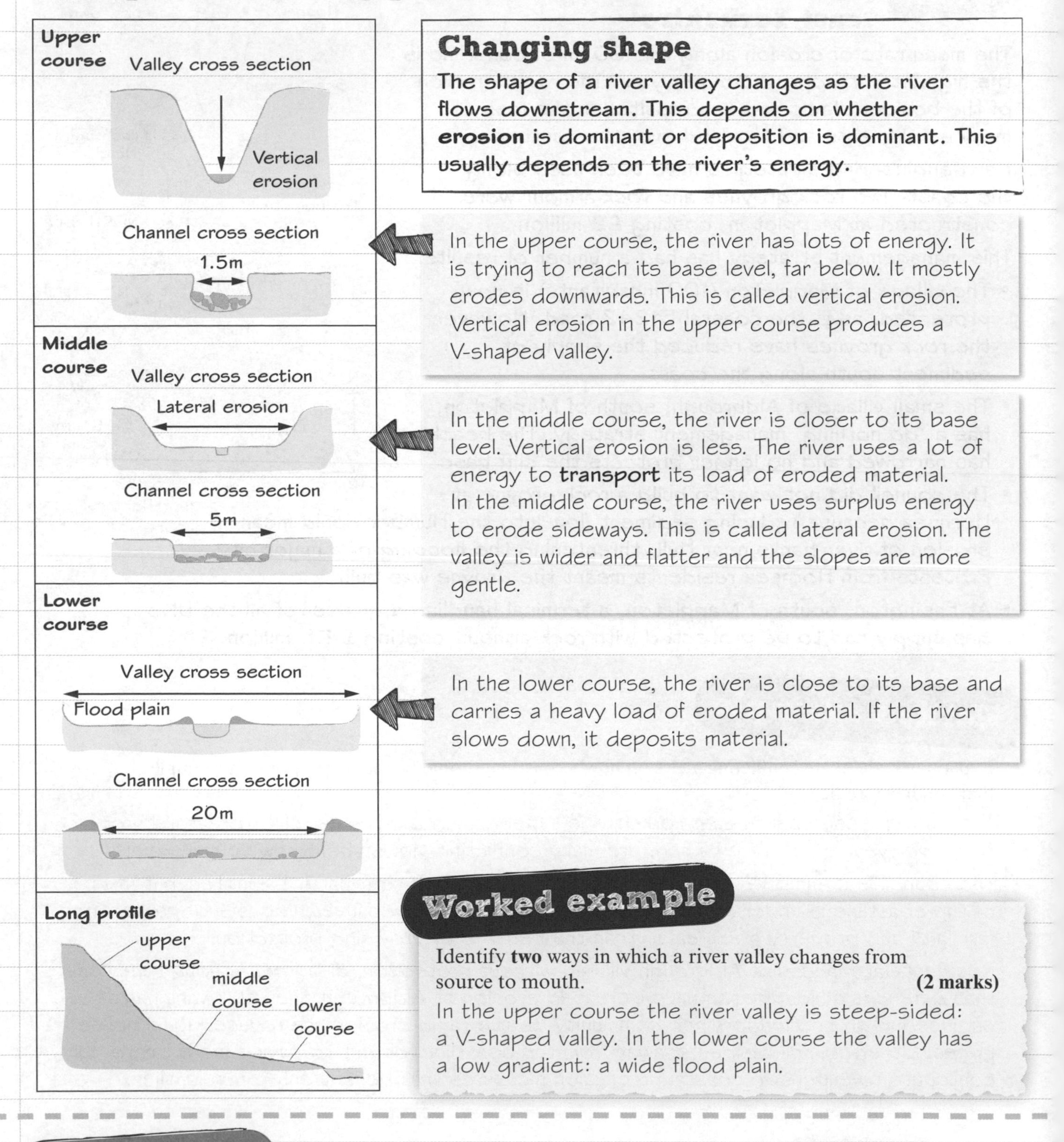

Changing shape

The shape of a river valley changes as the river flows downstream. This depends on whether **erosion** is dominant or deposition is dominant. This usually depends on the river's energy.

In the upper course, the river has lots of energy. It is trying to reach its base level, far below. It mostly erodes downwards. This is called vertical erosion. Vertical erosion in the upper course produces a V-shaped valley.

In the middle course, the river is closer to its base level. Vertical erosion is less. The river uses a lot of energy to **transport** its load of eroded material. In the middle course, the river uses surplus energy to erode sideways. This is called lateral erosion. The valley is wider and flatter and the slopes are more gentle.

In the lower course, the river is close to its base and carries a heavy load of eroded material. If the river slows down, it deposits material.

Worked example

Identify **two** ways in which a river valley changes from source to mouth. **(2 marks)**

In the upper course the river valley is steep-sided: a V-shaped valley. In the lower course the valley has a low gradient: a wide flood plain.

Now try this

Explain the difference between a cross profile of a river and a long profile of a river. **(2 marks)**

River processes

Fluvial processes – river processes – involve erosion, transportation and deposition. These processes impact on river landforms.

Processes of erosion

The four main processes of erosion

Hydraulic action
The force of the water on the bed and banks of the river removes material.

Attrition
The load that is carried by the river bumps together and wears down into smaller, smoother pieces.

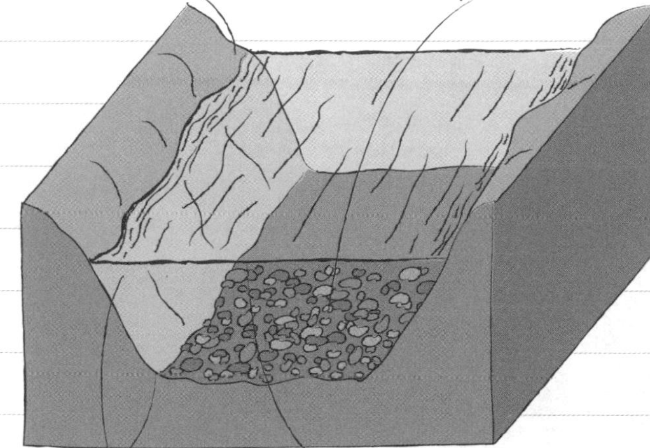

Abrasion
Material carried by the river rubs against the bed and banks and wears them away.

Solution
Some rock minerals dissolve in river water (e.g. calcium carbonate in limestone).

Worked example

Study this photo of the upper course of a river in Snowdonia. Explain why vertical erosion is more significant than lateral erosion here. **(2 marks)**

In the upper course, the river is far above base level (sea level) and erosion (mainly hydraulic action) is mostly vertical, deepening the river channel.

Lateral (sideways) erosion, takes place in the lower course, widening the river channel.

Transportation

The four main types of transportation. Transportation is the way in which the river carries eroded material.

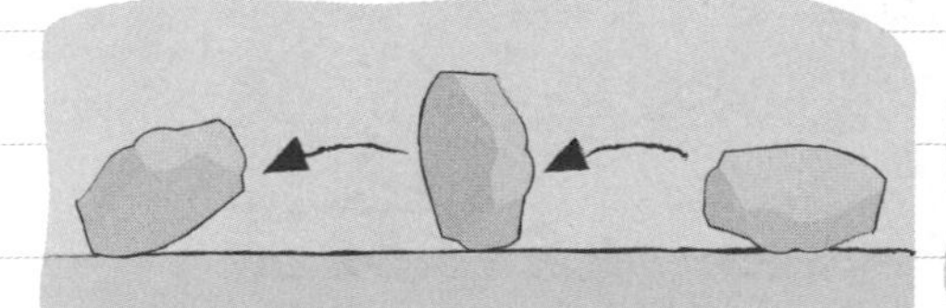

Traction: large boulders roll along the river bed

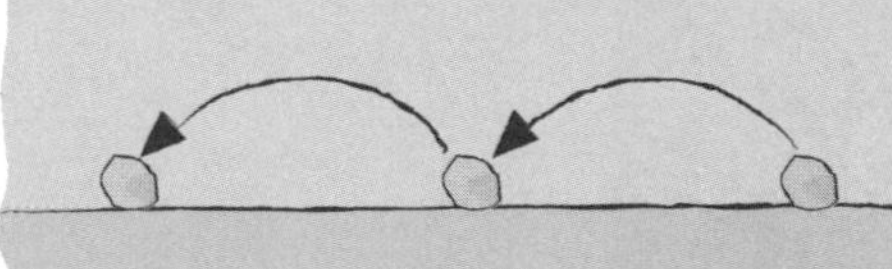

Saltation: smaller pebbles are bounced along the river bed, picked up and then dropped as the flow of the river changes

Suspension: finer sand and silt particles are carried along in the flow, giving the river a brown appearance

Solution: minerals from rocks such as limestone and chalk are dissolved in the water and carried along in the flow, although they cannot be seen

When the river loses energy (slows down) it may drop some of its load. This is called **deposition**.

Now try this

Identify **one** reason why rivers deposit sediment. **(1 mark)**

Erosion landforms

You need to know the characteristics of river erosion landforms, and how they are formed. The landforms of erosion are: interlocking spurs, waterfalls and gorges.

Waterfalls

A waterfall is a steep drop in a river's course. The diagram explains how they are formed.

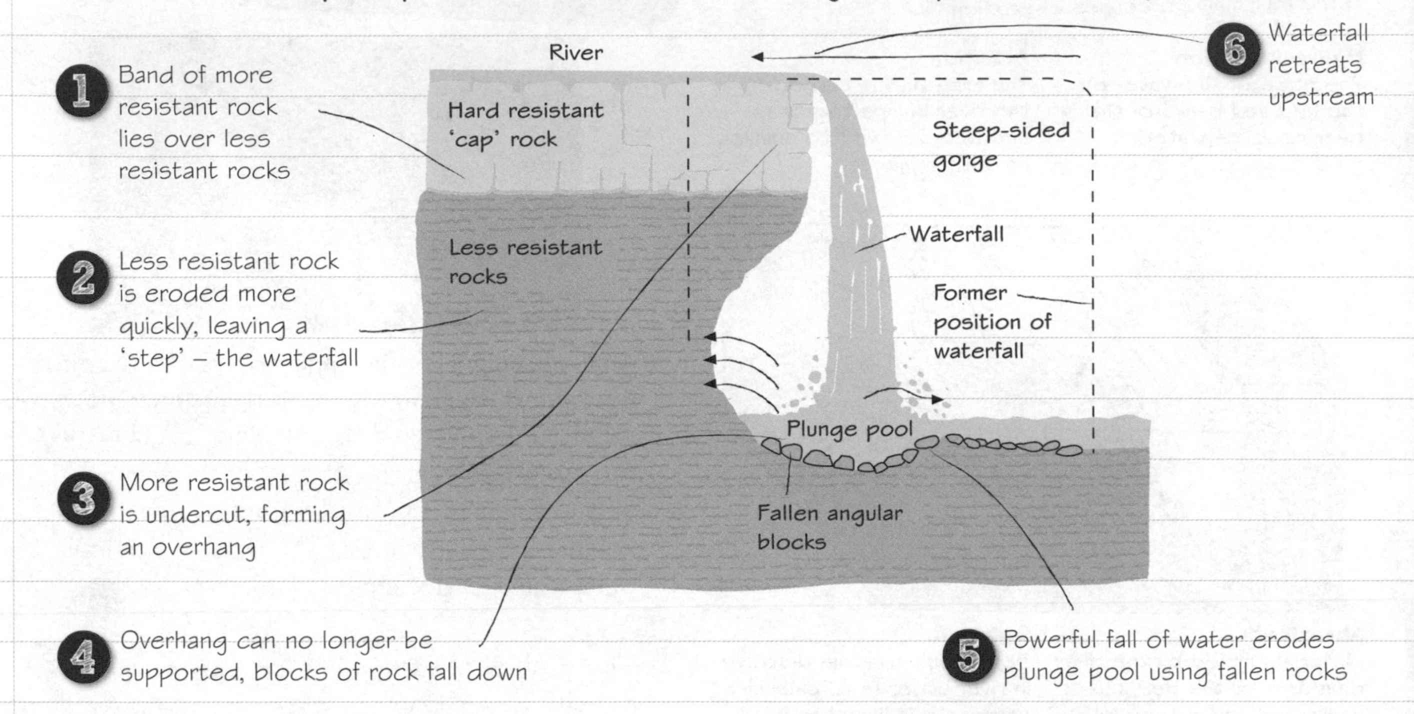

Gorges

Over a very long time, the process of undercutting and collapse is repeated and repeated, and the waterfall retreats upstream. A steep-sided gorge is formed.

Rivers have less power in the upper course: this is key for understanding landforms of the upper course.

Worked example

Study the diagram, which shows interlocking spurs. Explain how interlocking spurs form. **(2 marks)**

In the upper course of a river, near its source, the river has little power to erode rocks. Therefore it flows around the harder, more resistant rock. This creates spurs that interlock on either side of the valley as the river moves downstream.

Now try this

Study the diagram of a waterfall above. Identify **two** characteristics shown in the diagram that can also be seen in the picture of the gorge. **(2 marks)**

Erosion and deposition landforms

You need to know the characteristics of river landforms that result from both erosion and deposition, and how they are formed. These landforms are meanders and ox-bow lakes.

Meanders

In a meander, the river swings from side to side. The force of the water swinging from side to side directs the fastest current and greatest force of water against the outside bank of the meander, forming a steep bank called a **river cliff**. On the inside edge of the meander, the current is slower and deposition of sand takes place, creating a gently sloping bank called a **slip-off slope**.

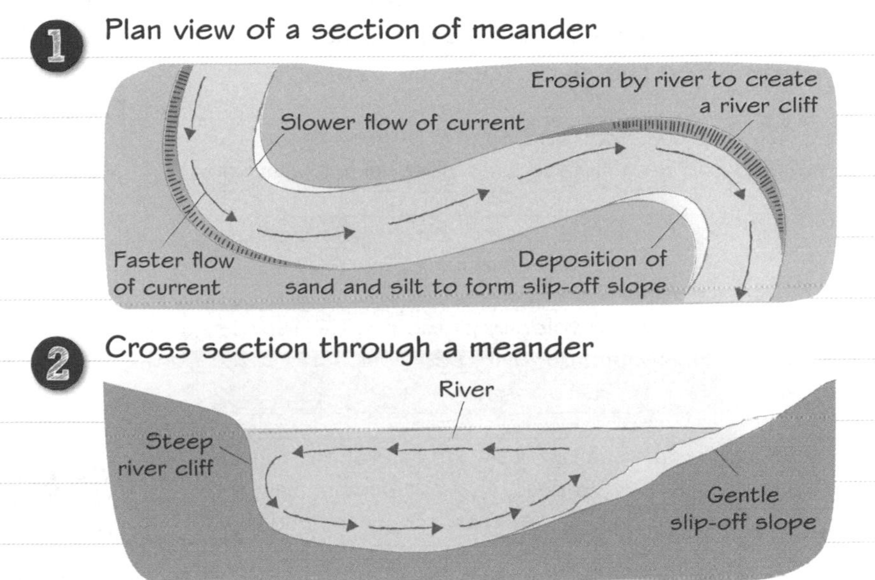

Ox-bow lakes

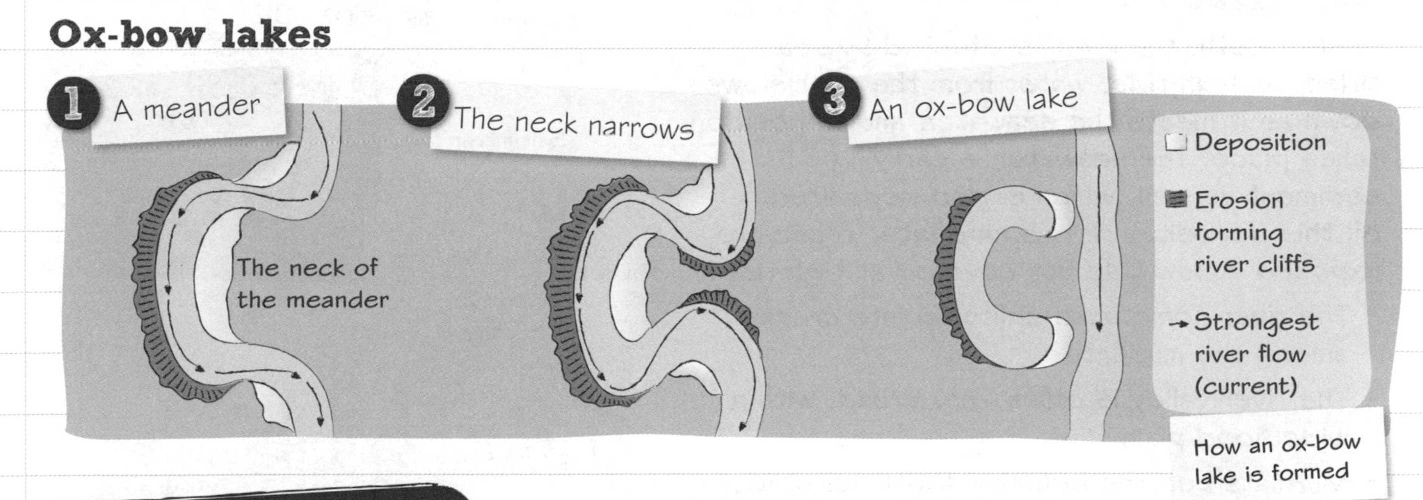

How an ox-bow lake is formed

Worked example

Study the 1:50 000 OS map extract of the River Towy, Wales. Describe how the processes of erosion and deposition are involved in forming the land form marked **A**. **(2 marks)**

Lateral erosion is greatest on the outside of a meander bend and forces the neck of the meander to narrow. Over time, the meander loop widens and the neck gets narrower, eventually eroding through. The meander is removed from the river current, creating an ox-bow lake.

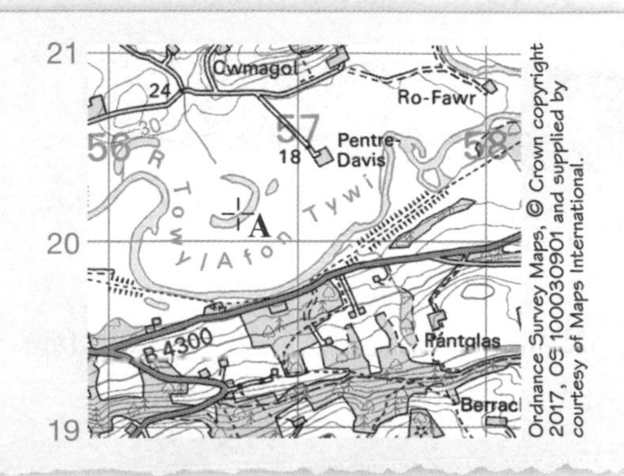

Now try this

Give the 6-figure grid reference for the ox-bow lake on the map extract above. **(1 mark)**

Had a look ☐ Nearly there ☐ Nailed it! ☐

Deposition landforms

You need to know the characteristics of river landforms that result from deposition, and how they are formed. These landforms are levées, flood plains and estuaries.

Flood plains

A flood plain is the wide, flat area of land either side of a river and experiences floods when the river tops its banks.

In the lower course, the river is nearing the sea and carries a huge amount of sediment (**alluvium**).

When the river floods, excess water spills over the surrounding area.

During flooding, the **velocity** of the river is reduced, it loses energy, and deposits sediment, forming the flood plain.

The flood plain is shaped by the **lateral** erosion of meanders as they gradually migrate downstream and by deposition of material on the inner bends.

How levées develop

The deposition process, which takes place during flooding, continues until eventually embankments are created beside the river. These are called **levées.**

Before a flood

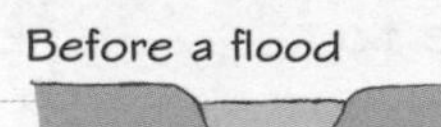

During a flood

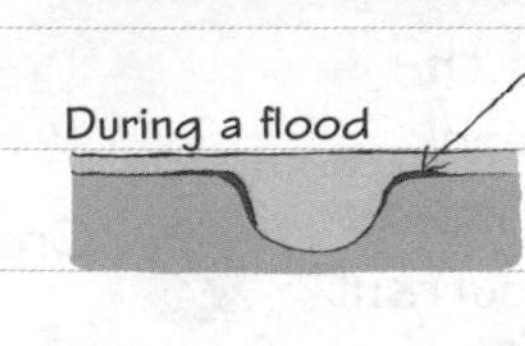

When flooding occurs, the heaviest material is deposited first due to the decrease in the river's energy. This material creates natural embankments called levées.

After a flood

The smaller and finer sediment, or alluvium, is deposited further from the river because it requires less energy to carry it.

Estuaries

At its mouth, the river is affected by sea tides. At high tide, water from the river slows down as it meets the seawater, and deposition takes place. The seawater is carrying sediment as well, which is also deposited. All this deposition forms **mudflats**. These are exposed at low tide but covered at high tide.

- The river sometimes splits up into creeks among the mudflats.
- The river valley is often very broad, with a wide flood plain.
- Mudflats become colonised with vegetation to form salt marshes.

Estuary of the River Ribble, UK

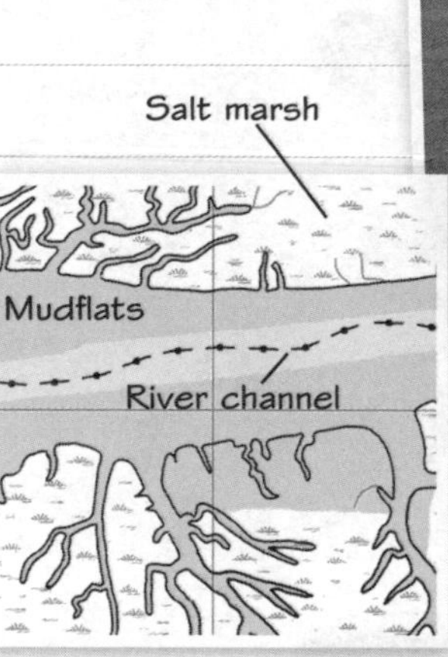

Worked example

Study the diagram below, which shows a cross section through a river channel.
Identify the landforms **X** and **Y**. **(2 marks)**

X: Levées **Y:** Flood plain

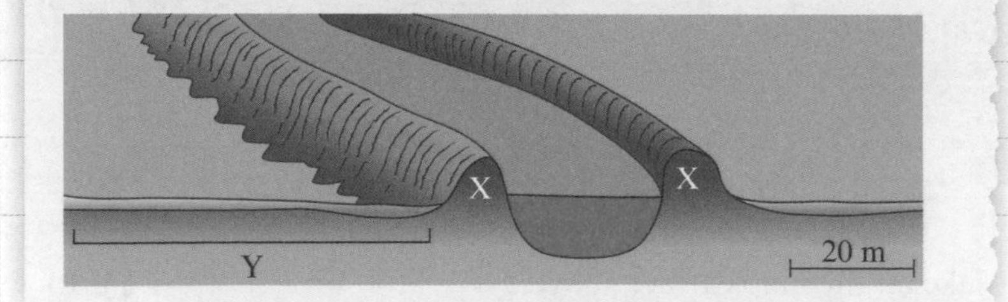

Now try this

Study the diagram opposite: the cross section through a river channel.
Explain how the landforms marked **X** are formed. **(4 marks)**

Remember to use specialist terms in your answer where you can: for example alluvium, embankment, velocity. Specialist terms demonstrate your knowledge and understanding, because they allow you to give very precise answers.

River landforms

You will have studied a UK river valley and identified landforms of erosion and deposition along it. **Make sure you revise the example that you studied in class.**

Example The River Conwy is 43 km long, beginning in the moorland plateau of Migneint Moor and ending at Conwy in North Wales.

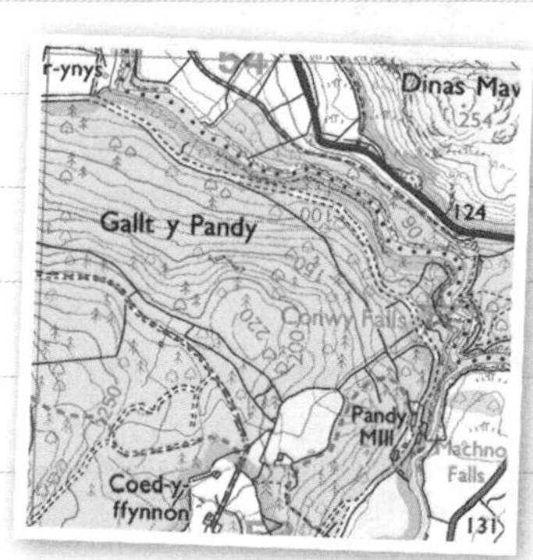

Upper course

Note the interlocking spurs and steep, V-shaped valley. There is a series of waterfalls here, where the level of the river drops 40 m in height.

What upper course landforms are there in your river valley example?

Worked example

Suggest **one** reason for the winding of the river channel by the Conwy Falls. **(1 mark)**

Interlocking spurs

The student has used the contour lines to work out that interlocking spurs is a better answer than meanders. Do you know why?

Middle course

This 1:25 000 OS map extract shows part of the middle course of the River Conwy. The river has formed a flat, wide flood plain here as it meanders through the valley. The dotted lines either side of the river are a mix of levées and artificial flood defences.

What middle course landforms are there in your river valley example?

Worked example

Explain **one** reason why flood plains are often used for farming. **(2 marks)**

Alluvium deposited on flood plains makes for very fertile soil as it contains many minerals.

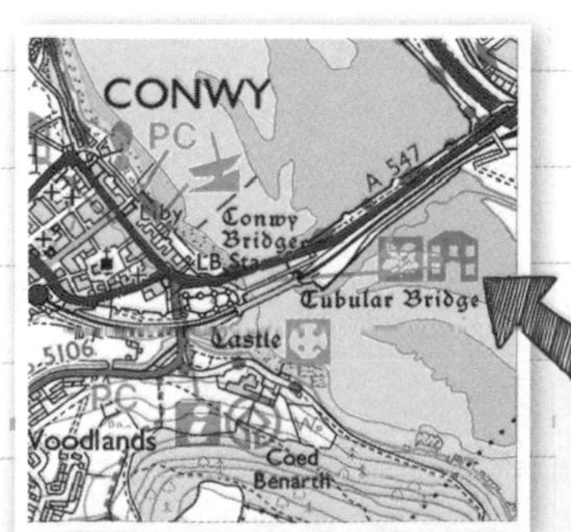

Lower course

This 1:25 000 OS map extract shows part of the River Conwy's estuary, where the river is tidal. When the tide is low, large expanses of mudflats are exposed.

What lower course landforms are there in your river valley example?

Now try this

There are nine different blue Ordnance Survey tourist and leisure symbols on the map extract of the Conwy river estuary. Identify **two** of them. **(2 marks)**

Had a look ☐ Nearly there ☐ Nailed it! ☐

Flood risk factors

River flooding occurs when the volume of water discharge is so great that all the water cannot be contained in the river channel. Different physical and human factors affect the risk of river flooding. Hydrographs show the relationship between precipitation and discharge.

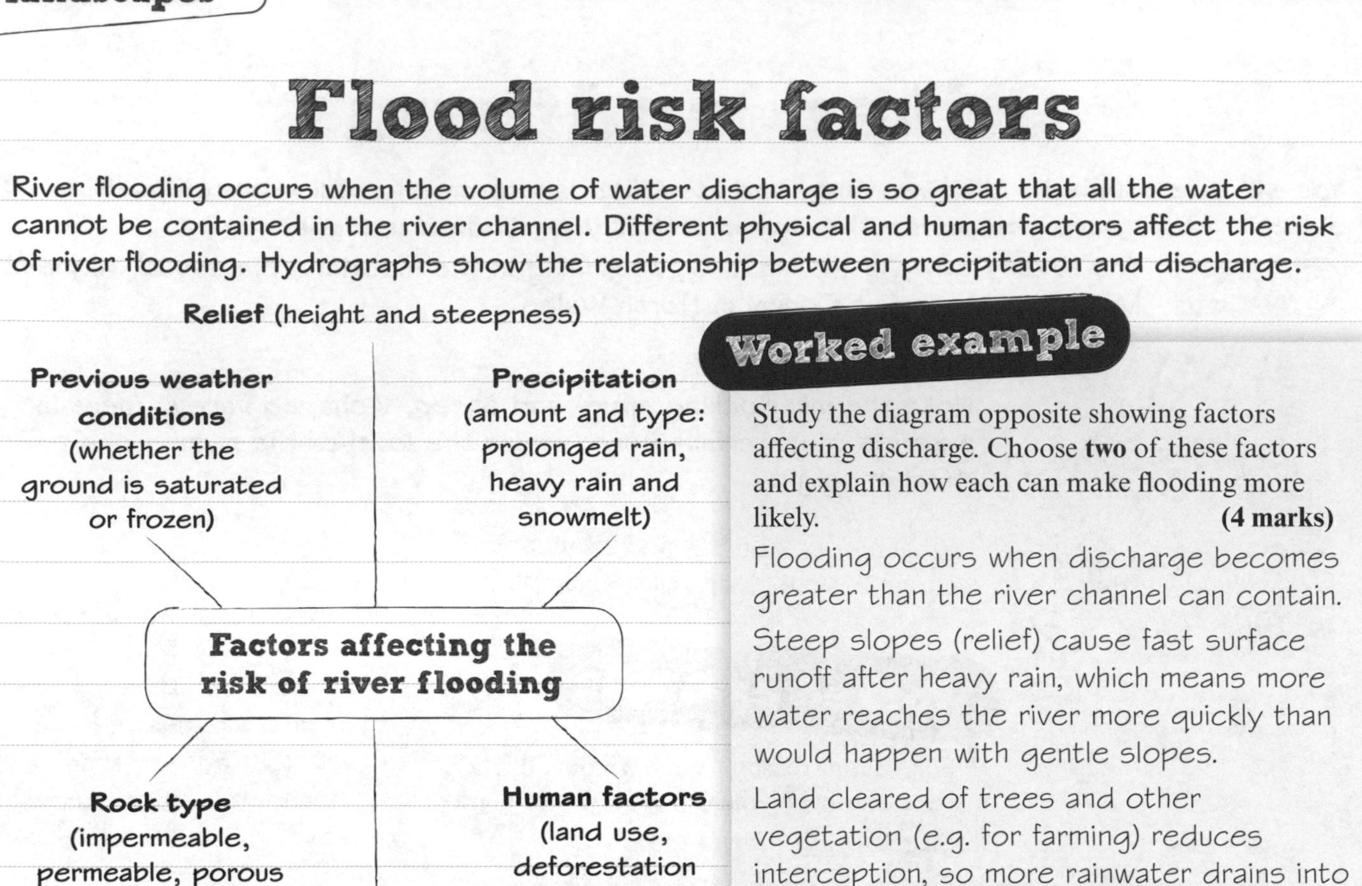

Worked example

Study the diagram opposite showing factors affecting discharge. Choose **two** of these factors and explain how each can make flooding more likely. **(4 marks)**

Flooding occurs when discharge becomes greater than the river channel can contain.

Steep slopes (relief) cause fast surface runoff after heavy rain, which means more water reaches the river more quickly than would happen with gentle slopes.

Land cleared of trees and other vegetation (e.g. for farming) reduces interception, so more rainwater drains into the river than would be the case if the slopes were heavily vegetated.

Flood hydrograph

A flood hydrograph (or storm hydrograph) shows how a river responds to a rainstorm.

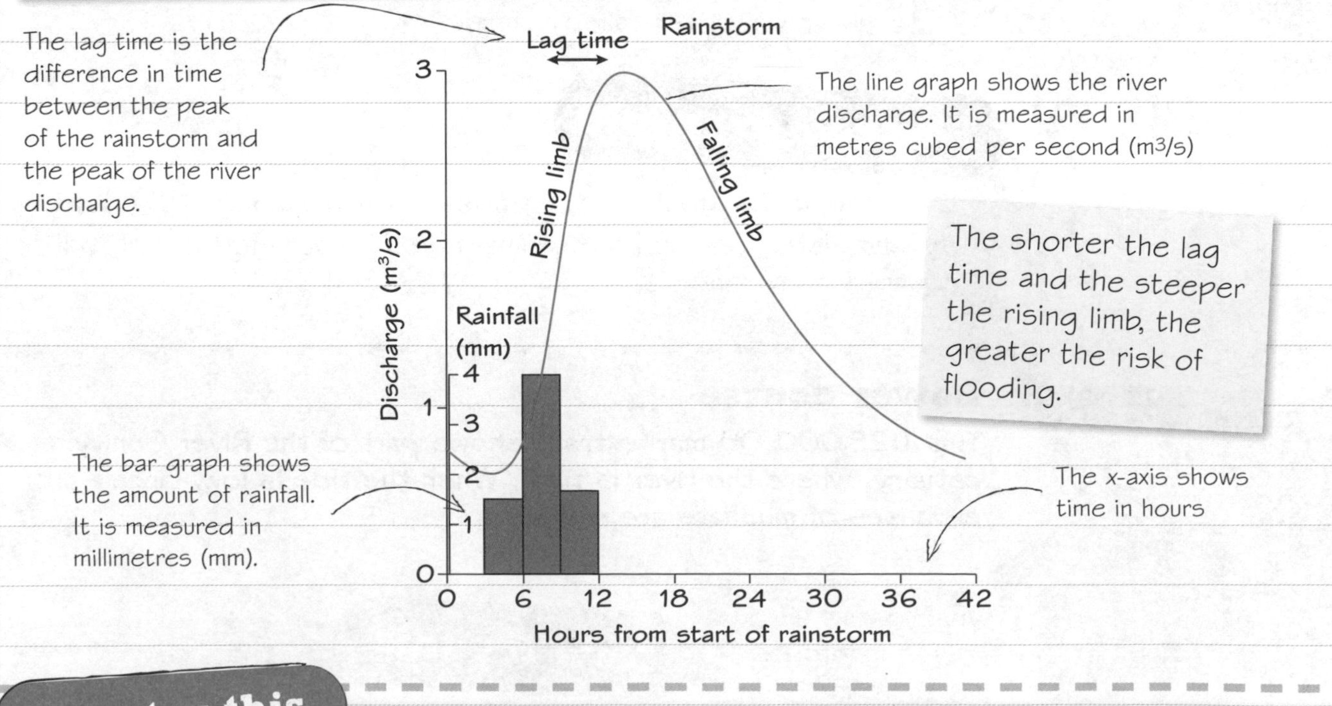

The shorter the lag time and the steeper the rising limb, the greater the risk of flooding.

Now try this

Study the diagram at the top of the page. Explain how precipitation and geology (rock type) can affect the risk of river flooding. **(4 marks)**

Hard engineering

You need to know about the costs and benefits of hard-engineering strategies to protect river landscapes from flooding: dams and reservoirs, straightening, embankments, flood relief channels.

Dams and reservoirs

Dams and reservoirs reduce the risk of flooding by storing water when rainfall is high then releasing it gradually through the year.

- 👍 Can also provide other benefits, such as HEP generation, tourist attraction, provision of drinking water
- 👎 Very expensive to build (mega-projects); need large areas of land to be flooded, so people may have to move house

River straightening

River straightening: converting a section of meandering river into a straighter (and often wider, deeper course) which reduces flooding by reducing channel friction, so water moves out of the area faster.

- 👍 Most done in the UK when river transport was economically important; straighter courses are quicker for boats to use
- 👎 Moving water quickly out of one area increases the risk of serious flooding downstream; straightened banks often collapse

Embankments

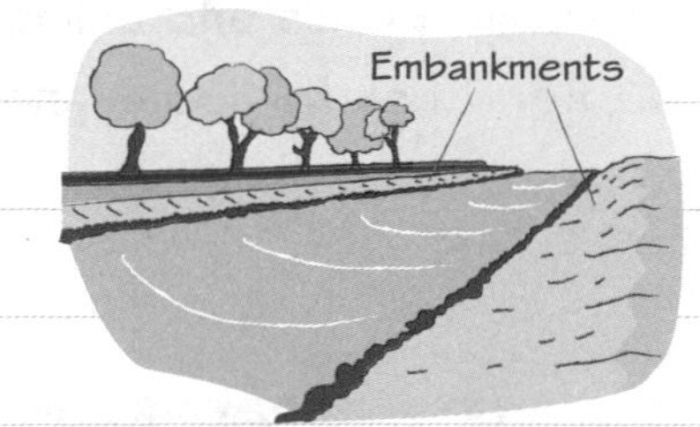

Embankments are artificially raised river banks that reduce flooding because the river can hold more water.

- 👍 Cheaper to build than other hard-engineering strategies – they can be simple earth banks piled up by bulldozer
- 👎 Less reliable than other strategies because, unless the embankments are repaired regularly, erosion weakens them

Flood relief channels

Flood relief channels are artificial river channels; when a river is flooding, water can be diverted into these channels to protect nearby housing.

- 👍 Very effective: a flood relief channel around a town gives thousands of houses a high level of protection from floods
- 👎 Increased risk of flooding downstream, where the relief channel rejoins the river; channels are also expensive

Now try this

The photo above right shows a flood relief channel at Tamworth during floods in May 2013.
Identify **two** benefits of the flood relief channel for residents of the houses in the photo. **(2 marks)**

Had a look ☐ Nearly there ☐ Nailed it! ☐

Soft engineering

You need to know about the costs and benefits of soft-engineering strategies to protect river landscapes from flooding: flood warnings and preparation, flood plain zoning, planting trees and river restoration.

Flood warnings and preparation

By constantly monitoring river levels and rainfall, the Environment Agency can warn people when a flood is likely, so they can prepare: for example, move furniture upstairs or install air vent covers.

- 👍 Warnings are cheap and effective; people can arrange for text messages or use social media to alert them
- 👎 Does not prevent flooding

Flood plain zoning

This restricts the sort of land use that is allowed on areas that are at high risk of flooding. Car parks and playing fields would be allowed; houses, schools and hospitals would not.

- 👍 Makes sure floods cause the minimum disruption to people and property
- 👎 No good for at-risk areas that have already been built on

Planting trees

Trees help reduce flooding risk by increasing infiltration and reducing surface runoff.

- 👍 Also reduces soil erosion, especially on slopes; can enhance the look of area
- 👎 Land planted with trees may not also be usable for other uses, such as housing, farming; trees take time to grow and take effect

River restoration

This includes returning straightened rivers to their natural courses, removing concrete channels, reintroducing marshland and lowering flood plains. These measures slow river flow, reducing flood risk downstream.

- 👍 Restoring natural flood plains creates habitats for wild plants and animals
- 👎 Increased flood risk for local land users

Worked example

Study the photo of people using sand bags and a temporary plastic barrier to contain flood water.

(a) Which one of the following soft-engineering strategies is this plastic barrier an example of? **(1 mark)**

☒ Flood warnings and preparation ☐ Planting trees
☐ Flood plain zoning ☐ River restoration

(b) Suggest **one** economic impact that frequent river flooding would have on residents in the area shown in the photo. **(2 marks)**

To protect their property from damage, residents of this area might buy their own sand bags and plastic barrier to use as a community when the river was about to flood. Buying these products would have an economic impact, but less impact than if their houses flooded.

Now try this

The Environment Agency has a web page to warn people about flooding. Suggest **one** problem with providing flood warnings using websites and social media. **(2 marks)**

Flood management

You need to know an **example** of a flood management scheme to use in questions about protecting river landscapes from the effects of flooding.

Why the scheme was required

Flooding affects many areas, so what was important enough about your example area for there to be investment?

- Were a large number of houses at risk?
- Had a severe flood caused damage, leading to pressure on the government to better protect people from floods in the future?
- Did flooding threaten high-value infrastructure, such as power stations?

Flood management scheme

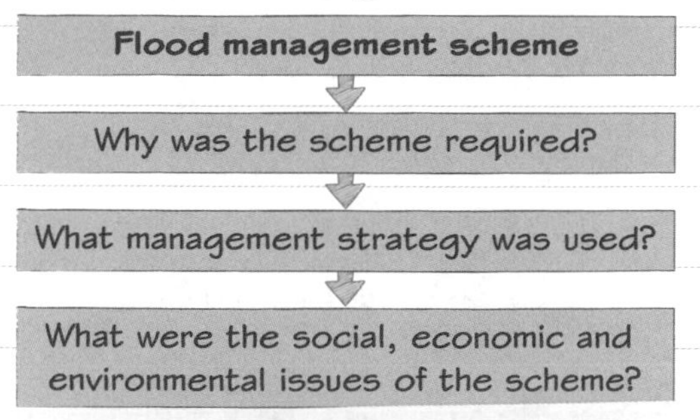

The information you need for your flood management scheme example.

The management strategy

You need to know the management strategy used in your example.

The management strategy can be hard engineering or soft engineering or a mixture of both. Useful details to know include:

- how the area was protected in the past
- what the strategy was designed to do
- the cost of the strategy
- when it was started and completed
- how successful it has been at preventing floods since it was completed.

A 'leaky dam' in Pickering, Yorkshire

Example A concrete wall to protect Pickering would have cost £20 million, which was not affordable. Instead, soft-engineering measures slow the river down, make use of flood plain storage and increase infiltration.

Issues

You need to know what social, economic and environmental issues have come from your chosen flood management scheme.

- **Social issues** – for example, some areas are protected, while others are not.
- **Economic issues** – schemes can be expensive, but may not work as well as expected; the cost of defending some areas is considered too high, or too high for hard engineering.
- **Environmental issues** – issues can be positive as well as negative: for example, soft engineering can create new habitats.

Worked example

Using an example of a UK flood management scheme you have studied, explain an economic issue resulting from the scheme. **(4 marks)**

Flood management schemes have to balance cost against expected flood risk. In Cockermouth, devastating floods in 2009 meant that a new self-raising flood barrier was installed at a cost of £4.4 million. A decision was made about the maximum height of the barrier based on cost, technical issues and predictions about future flood risks. Unfortunately, floods in 2015 overtopped the barrier after record rainfall. Local residents felt the money had been wasted.

Now try this

Describe **one** social issue, **one** economic issue and **one** environmental issue from your example. **(3 marks)**

Had a look ☐ Nearly there ☐ Nailed it! ☐

Glacial processes

Much of the UK's landscape was once covered by ice, and most of the UK's landscape was shaped by glacial processes. **Only revise glacial landscapes if you did them in class.**

Freeze–thaw weathering

Freeze–thaw weathering is very active in glaciated areas

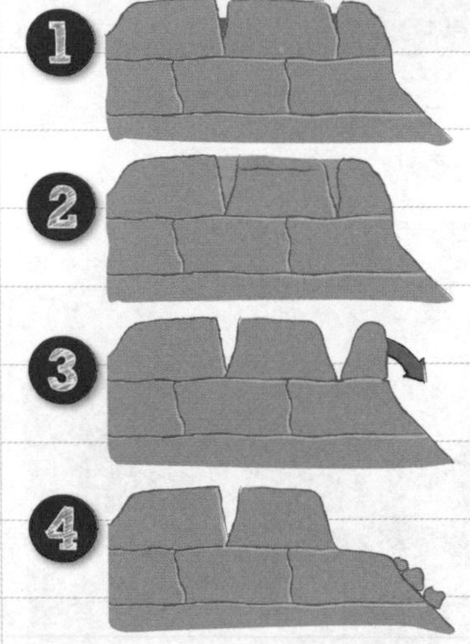

1. Water fills a crack or joint in the rock.
2. Water freezes and the crack is widened.
3. Repeated freeze–thaw action increases the size of the crack until the block of rock breaks off.
4. Loose blocks of rock are called scree.

Glacial erosion

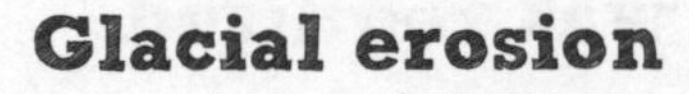

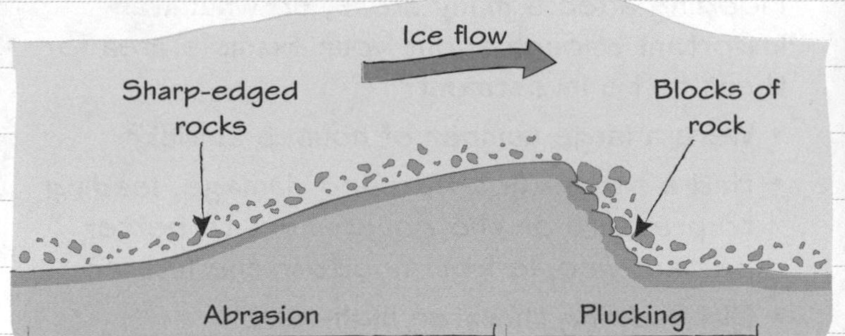

Small, sharp-edged rocks and rock particles embedded in the bottom of the glacier wear away and polish the bedrock the glacier is passing over. This is called **abrasion**. Larger rock fragments scrape the bedrock causing **striations**.

Blocks of bedrock freeze to the bottom of the glacier and get plucked out as the glacier moves down the valley. This is called **plucking**.

Transportation

Glaciers are powerful and erode and transport a huge amount of material, known as **moraine**. Moraine is transported on top, within and underneath the glacier.

- **Bulldozing** happens when a glacier retreats and then moves forward again, pushing debris ahead of it.
- **Rotational slip** happens in summer, when a thin layer of meltwater beneath the glacier means it sometimes slides forward. In a depression, this slide can have a curving motion.

Deposition

Most deposition happens when the ice melts. Since ice melts most at the **snout** (the furthest point of the glacier), this is where most deposition happens. Deposition also happens if the ice thins or goes around an obstacle so it cannot carry as much load.

Worked example

This map shows the extent of ice cover in the UK in the last glaciation, which ended 12 000 years ago.

Briefly describe the main impacts that glaciation had on the UK landscape. **(4 marks)**

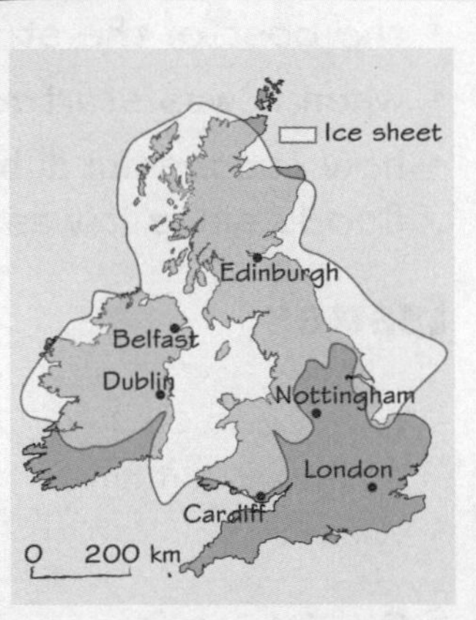

The top half of the UK was glaciated. Ice sheets and glaciers hundreds of metres thick pressed down on the landscape and eroded it in distinctive ways. The bottom half of the UK was not glaciated, but was heavily influenced by glacial deposition. Clays, sands and silts eroded by glaciers were dumped and washed over southern areas.

Now try this

Study the diagram of glacial erosion on this page. Explain why abrasion sometimes produces smoothing of the bedrock under the glacier, and sometimes striations. **(4 marks)**

Erosion landforms 1

Distinctive landforms result from different processes. Glacial erosion produces corries, arêtes and pyramidal peaks.

A corrie

Freeze–thaw weathering supplies rocks for abrasion.

Water seeps down the bergschrund crevasse, which helps rotational slip and means the headwall is also affected by freeze–thaw.

Scree

Bergschrund (large crevasse)

Crevasses

Ice movement

Rotational slip

Rotational slip puts more pressure on the bottom of the hollow than where the glacier leaves the hollow.

A lip forms where erosion is less powerful.

Ice sticking to the headwall plucks blocks of rock away from it.

Abrasion scoops out the bottom of the hollow.

How a corrie is formed

An arête

An arête forms when two corries form back to back

A pyramidal peak

A pyramidal peak forms when three corries cut backwards

Worked example

Study the OS map extract opposite (1:50 000 scale) showing Snowdon, a mountain in Wales.
Which of the following features is at grid reference 610544?
(1 mark)

- ☐ A corrie
- ☒ A pyramidal peak
- ☐ A U-shaped valley
- ☐ A hanging valley

Now try this

What feature of this glaciated landscape is found in the corrie hollow at 617546? **(1 mark)**

Erosion landforms 2

Glacial erosion also produces truncated spurs, glacial troughs, ribbon lakes and hanging valleys.

A valley before glaciation

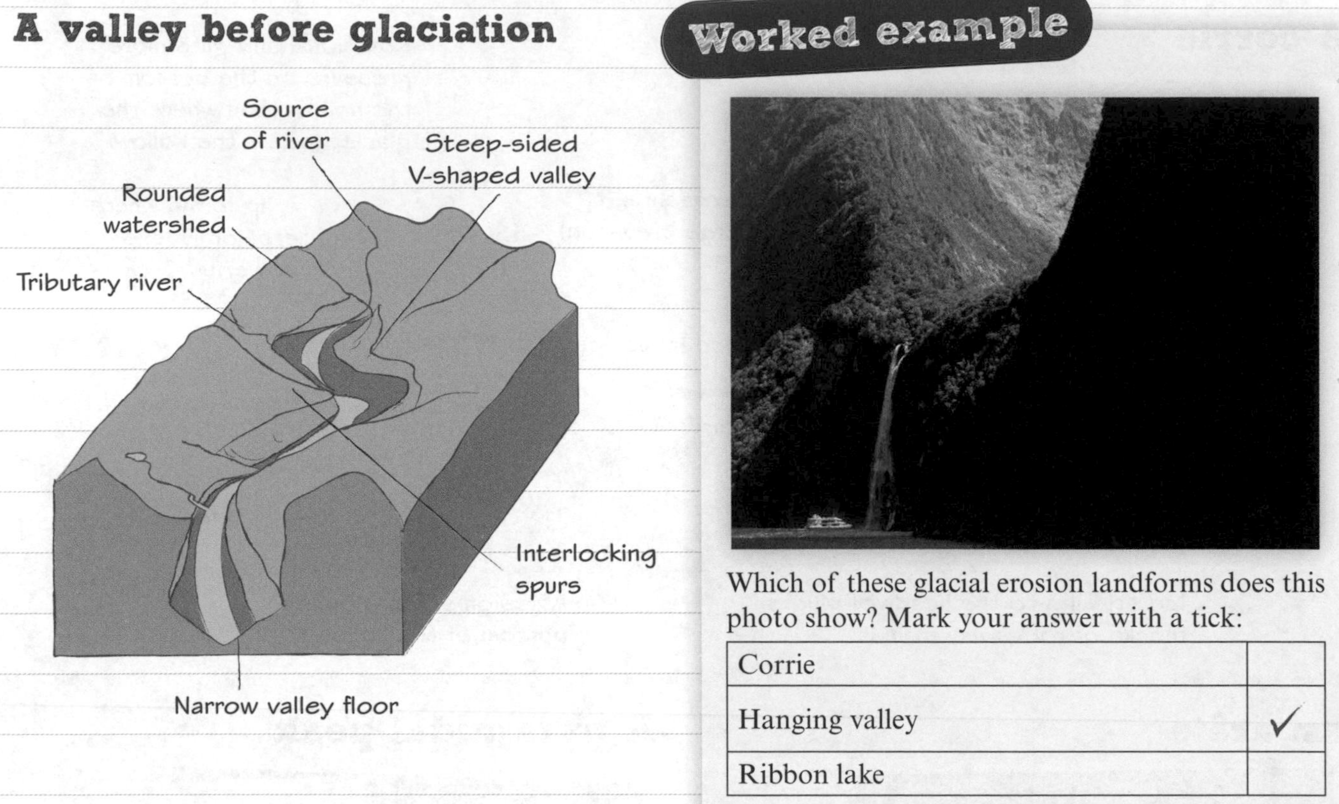

Worked example

Which of these glacial erosion landforms does this photo show? Mark your answer with a tick:

Corrie	
Hanging valley	✓
Ribbon lake	

(1 mark)

A valley after glaciation

Jagged watershed

Truncated spurs are formed when the glacier pushes straight through the valley and removes the ends of the interlocking spurs of the old river valley.

Tributary in a hanging valley with a waterfall. **Hanging valleys** are old tributary valleys, occupied by small glaciers that can't erode down as far as the main glacier.

Ribbon lakes are long and thin and form when a glacier erodes a long, thin section of the valley floor more deeply than the rest. This usually happens when the ice gets thicker where another glacier joins as a tributary or where a valley has been blocked by a terminal moraine and meltwater builds up behind.

Glacial trough is a glaciated valley. A glacier has deepened and changed the shape of a former river valley from a V shape to a U shape.

Check your diagram carefully to make sure everything is where it should be and all labelling is in the right place, with the head of the arrow actually **touching** the feature identified.

Now try this

Draw and annotate a cross section of a glaciated valley to include the following features:

U-shaped valley **Hanging valley** **Flat valley floor** **Waterfall** **(4 marks)**

Transportation and deposition landforms

Some glacial landforms are produced by a combination of transportation and deposition: different types of moraine, drumlins and erratics.

Moraines

Lateral moraine
A build-up of material along the side of a glacier. When the glacier melts this forms a ridge along the valley side.

Medial moraine
When a tributary glacier joins the main glacier, lateral moraines merge towards the middle of the main glacier. When the glacier melts the moraine is deposited in a ridge down the middle of the valley.

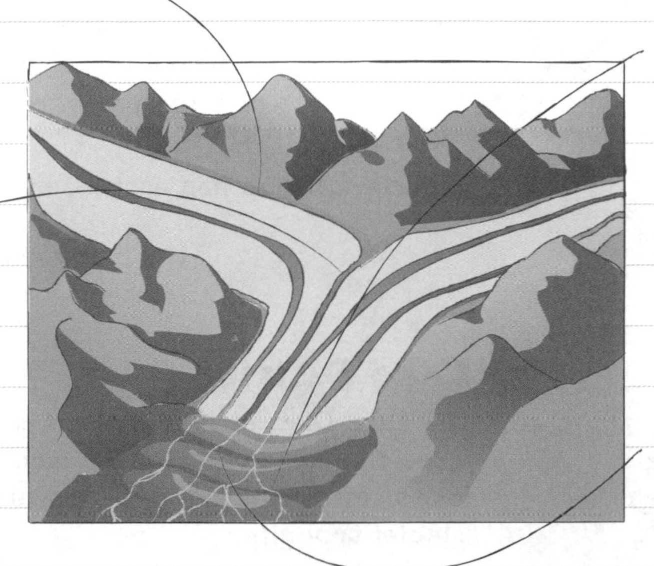

Terminal moraine
Most deposition occurs at the snout where great ridges of material pile up. Terminal moraines mark the furthest extent of the glacier (terminal means end).

Ground moraine
As the glacier melts it drops material known as boulder clay or till all over the valley floor, leaving hummocky ground.

Drumlins

Drumlins are low hills (30–40 metres high) that are blunt at one end and tapered at the other. The tapered end points in the direction of the glacial flow that produced the drumlin.

Drumlins form when melting ice is pushed forward over a lowland area, depositing as it goes. Where there are obstacles, glacial deposition is increased, making the blunt end of the drumlin. Further deposits are moulded into the tapered end as the glacier continues its course. Lots of drumlins can occur together, described as a 'basket of eggs' landscape or a drumlin 'swarm'.

Erratics

Erratics are large boulders deposited by glaciers. They are often different rock types than the bedrock onto which they are deposited.

Worked example

Identify the depositional landform in the photo above. **(1 mark)**

Drumlins

Now try this

Look at the diagram at the top of this page. Suggest **one** reason why there are two other ridges behind the terminal moraine. **(2 marks)**

Upland glaciated area

You will have studied an **example** of an upland area in the UK which has been affected by glaciation. You need to revise this example so you can use it to write about its major landforms of erosion and deposition. **Revise the example you studied in class, following the prompts included here.**

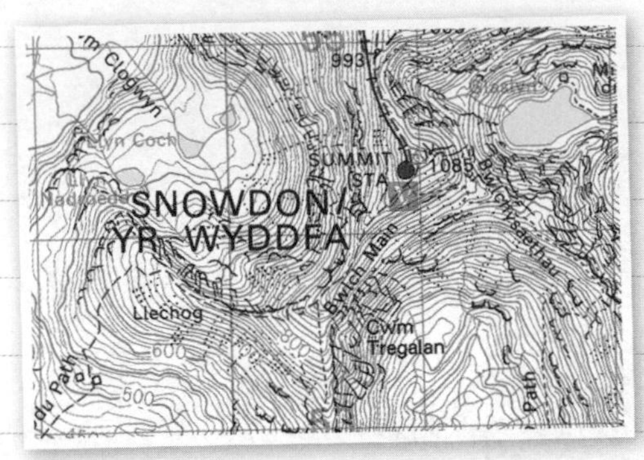

Ordnance Survey Maps, © Crown copyright 2017, OS 100030901 and supplied by courtesy of Maps International.

Example Snowdonia

Snowdonia is an upland glaciated area. Snowdon is 1085 m high and is the highest mountain in the UK outside Scotland. It is a **pyramidal peak**, with **corries** scooped out by glaciers on three sides.

- The corries now feature corrie lakes (tarns), such as Glaslyn (6154).
- Snowdon also has arêtes, such as Bwlch Main (6053).

What erosion landforms are there in your upland area example?

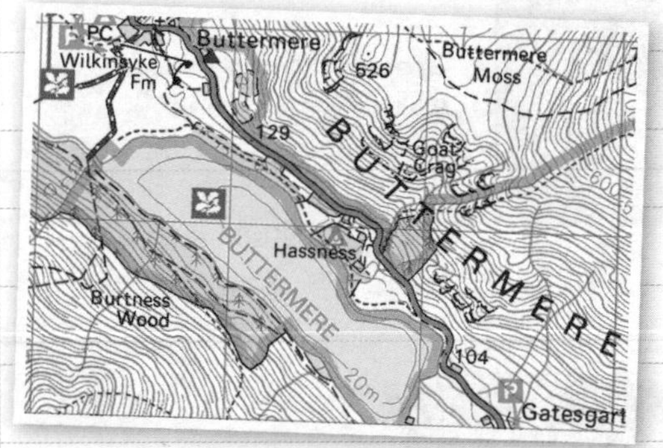

Ordnance Survey Maps, © Crown copyright 2017, OS 100030901 and supplied by courtesy of Maps International.

Lake District

The Lake District is an upland glaciated area. The lakes that give the area its name are often **ribbon lakes**, formed in deep, steep-sided, flat-bottomed **glacial troughs**.

Goat Crag (1916) is a truncated spur with a hanging valley.

Worked example

Study the photo, which shows the Bowder Stone in the Lake District. The Bowder Stone is probably originally from Scotland.

(a) What type of glacial landform is the Bowder Stone? **(1 mark)**

An erratic

(b) Outline the processes by which the Bowder Stone reached the Lake District. **(2 marks)**

This erratic was eroded from bedrock in Scotland, probably when the glacier undercut a rock cliff. It was then transported to the Lake District on the glacier and deposited there when the glacier ice melted.

Now try this

Study the photo of Goat Crag and the map extract of Buttermere. Which direction was the photographer facing when taking this picture of Goat Crag? **(1 mark)**

Activities and conflicts

The main economic activities in glaciated upland areas are tourism, farming, forestry and quarrying. Sometimes the different ways in which land is used are in conflict with each other.

Farming

Upland glaciated areas are tough areas to farm. The soils are often thin and acidic. Flat, fertile land is restricted to the valley bottoms.

Most of the land is too steep and exposed to the wind and cold winter temperatures for crops. The traditional land use for glaciated upland areas is sheep farming.

Profits are hard to make in upland farming, and most farming families have **diversified** into other economic activities: for example, setting aside fields for campsites.

Conservation

Where tourism is very popular, it can start to damage what makes the upland area attractive to visitors. For example:

- roads and car parks become jammed with visitors' cars and tourist coaches
- footpaths become eroded and fragile upland plant life is trampled
- litter contaminates local ecosystems.

Tourism

Glaciated upland areas have many advantages for tourists: their dramatic landscapes attract walkers, ribbon lakes are popular for watersports and angling, the steep slopes attract skiers in winter and mountain bikers and hill runners all year round. Tourism brings large numbers of people to some glaciated upland areas: 17.3 million people visited the Lake District in 2015, for example, bringing £1 billion into the area's economy.

Forestry

Forestry makes economic sense in upland areas where the land has low value and population numbers are low. Most plantations are conifers: quick-growing softwood trees.

Quarrying

The resistant rocks making up upland areas are in demand for building and road construction.

Conflicting land use

Some economic activity conflicts with other ways in which land in glaciated upland areas is used. For example:

- conifer plantations can block views for tourists, and look unsightly when felled
- blasting rocks in quarrying disturbs the peaceful environment that visitors enjoy
- valuable quarrying and mining in upland areas may not be permitted because of the impact on the environment and on tourism
- tourists forget to close farm gates, letting livestock out into fields of crops
- use of upland areas for windfarms to generate renewable energy conflicts with tourism if tourists think they are ugly.

Worked example

This photo shows footpath conservation in the Lake District following extensive footpath erosion. Suggest **one** other way in which tourism can have a negative impact on glaciated upland areas. **(1 mark)**

Watersports on lakes can cause erosion of the lake shoreline.

Now try this

Describe **two** ways in which the use of upland glaciated areas for tourism can cause conflicts with other land uses, such as farming, forestry or quarrying. **(2 marks)**

Tourism

You need an **example** of how a glaciated upland area in the UK is used for tourism to write answers about tourist attractions, impacts of tourism and strategies to manage those impacts. Here we look at the Lake District. **Make sure you revise the example you did in class.**

Attractions for tourists

Attractions

- Ribbon lakes for watersports and pleasure boats; rivers for canoeing
- Beautiful, dramatic scenery
- Historical links: famous poets, artists, writers
- Well-marked and signposted walks for casual walkers
- Shops selling local specialities, crafts and souvenirs
- Challenging mountain climbs and rock faces
- Opportunities for winter sports: skiing and snowboarding
- Adventure holidays: kayaking, mountain biking, pony trekking

What are the attractions for tourists in your example glaciated upland area?

Example

Impacts of tourism

Social impacts

- Tourists who buy second homes in the area make housing unaffordable for locals
- Jobs in tourism are often seasonal – may only last for the summer

Economic impacts

- Tourists bring money into the local economy: for example, £1 billion for the Lake District
- Tourism provides jobs for local people, though they are often low-paid

Environmental impacts

- Footpath erosion
- Trampling damage to plants and crops
- Litter and pollution

This photo shows a shop at Ullswater in the Lake District. Suggest **one** way in which tourism contributes to the economy of an upland glaciated area you have studied. **(2 marks)**

In the Lake District, 16 000 jobs are in tourism: it is the area's main source of income.

Managing the impact of tourism

Managing the impact

- Car and coach parking moved to sites at edge of towns
- Information for tourists on why they should take litter home
- Damaged areas fenced off so they can recover
- One-way systems and speed bumps introduced to calm traffic
- Tourists encouraged to contribute to conservation and restoration projects
- Zoning of lakes so conflicting uses do not overlap
- Public transport promoted with good bus service to popular locations
- Heavily used footpaths repaired with hard-wearing stones

What strategies are used to manage the impact of tourism in the area you studied?

Now try this

Describe **one** strategy used to manage the impact of tourism in the glaciated upland area that you have studied. **(2 marks)**

Global urban change

HICs (high-income countries) have slower urbanisation rates than **LICs** (low-income countries) and **NEEs** (newly emerging economies) because HICs are already highly urbanised. For example, 80 per cent of people in the UK already live in cities.

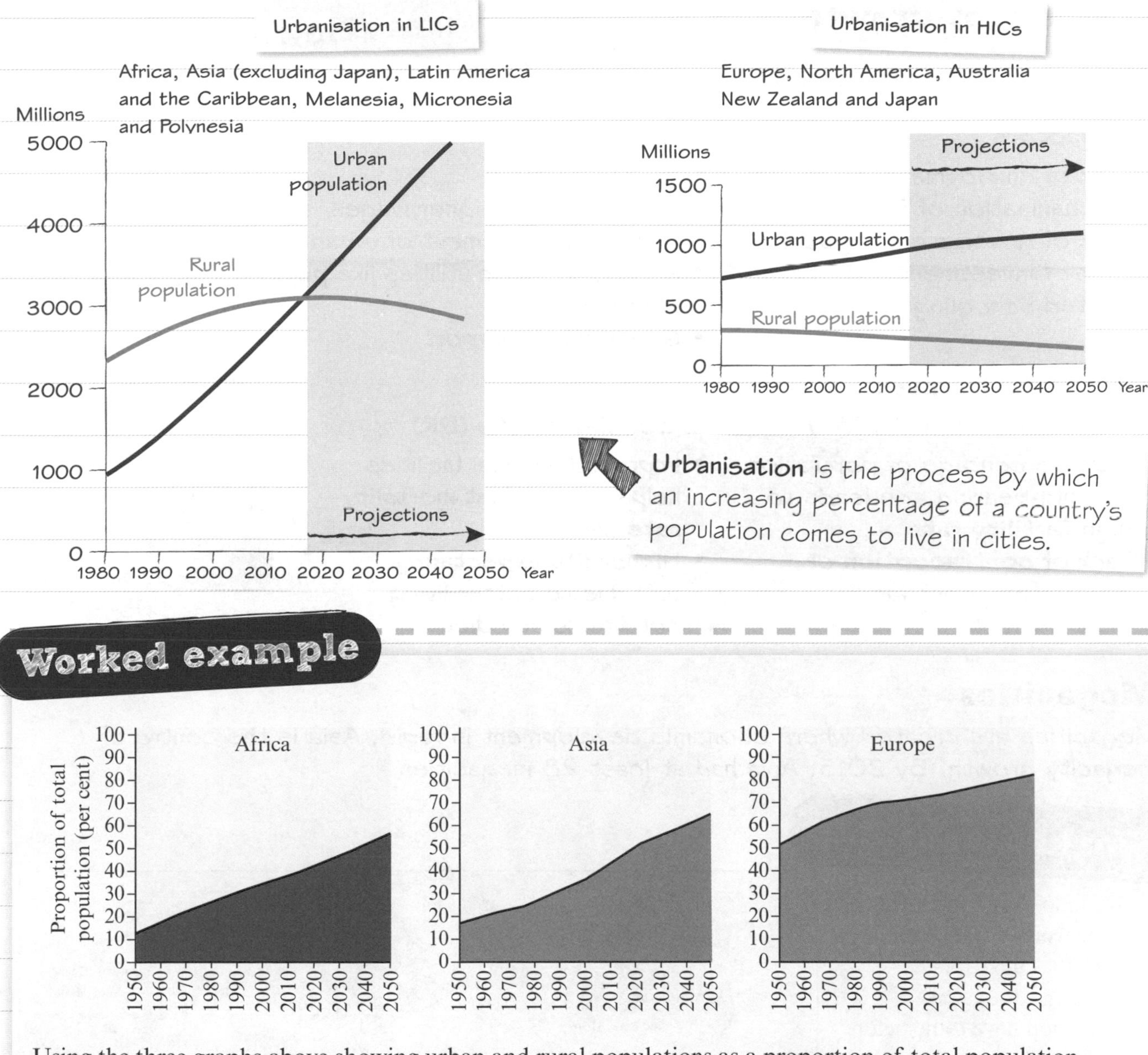

Urbanisation is the process by which an increasing percentage of a country's population comes to live in cities.

Worked example

Using the three graphs above showing urban and rural populations as a proportion of total population, suggest **two** reasons for the high rate of urban growth in many LICs. **(2 marks)**

Unlike HICs where the majority of people (73.4% in Europe) are already living in cities, in LICs in Africa and Asia most people still live in rural areas so there is a lot more potential for urban growth.

LIC rural areas usually have few opportunities and life is very tough (subsistence farming). LIC cities have many more opportunities (a much greater range of jobs, health and education opportunities) and offer a better quality of life.

Now try this

The world's urban population in 2000 was 2.84 billion. By 2005, it had risen to 3.15 billion. What was the percentage rate of change from 2000 to 2005? Round to the nearest whole number. **(1 mark)**

Urbanisation factors

The two main factors affecting the rate of urbanisation are migration and natural increase. **Megacities** have emerged in NEEs and LICs because of high rates of natural increase and migration from rural areas to urban areas.

Reasons for growth

1 Push–pull theory (rural to urban migration)

There are **rural push factors** (factors that 'force' you to leave an area) and **urban pull factors** (factors that 'encourage' you to move).

Push factors

- Lack of employment
- Mechanisation of farming
- Lack of government support
- Lack of investment
- Limited schooling

Pull factors

- Employment opportunities, higher wages
- Pace and excitement of urban life
- Access to public utilities like piped water, sewage and electricity
- Government support

2 Natural increase

Increase in birth rate (BR)

- High percentage of population of child-bearing age leads to high **fertility rate**
- Lack of **contraception** or **education** about family planning

Lower death rate (DR)

- Improved medical facilities helps lower **infant mortality rate** (IMR)
- Higher **life expectancy** (LE) due to better living conditions and diet

Natural increase is the rise in population caused by the birth rate exceeding the death rate. **Fertility rate** is the number of live births per 1000 women of child-bearing age (15–44).

Megacities

Megacities are created where economic development is rapid. Asia is the centre of megacity growth. By 2015, Asia had at least 28 megacities.

Worked example

This map shows the location of the world's 20 largest megacities in 2015.

How large does a city's population have to be before it is classed as a megacity? **(1 mark)**

10 million people

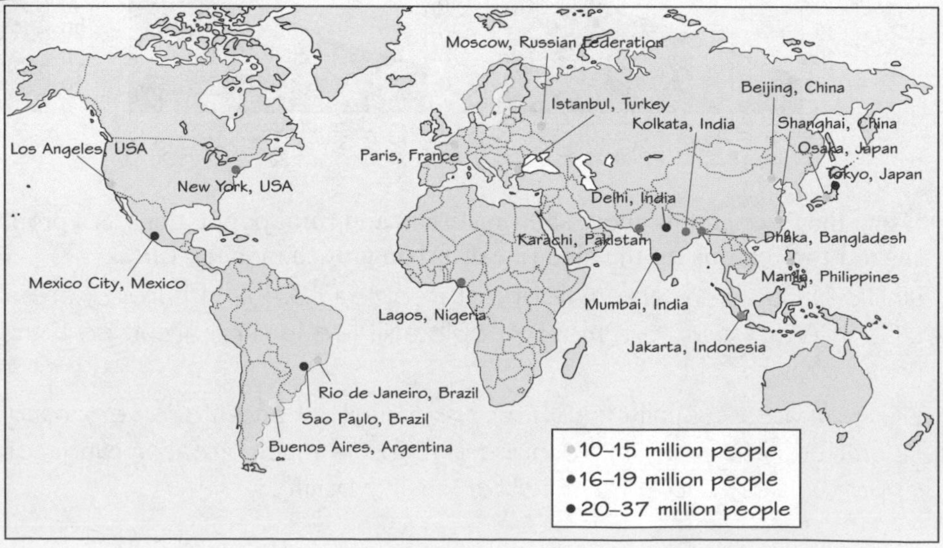

Now try this

In 2015, London had a population of 8.6 million. Calculate the percentage increase required for London to become a megacity. Round your answer up to one decimal place. **(1 mark)**

Non-UK city: location and growth

Case study You need to know a **case study** of a major city in an LIC or NEE to show the different ways urban growth creates opportunities and challenges in your city. Here we look at Mumbai. **You should revise the city that you studied in class, using the prompts here to help you.**

Location of your city

You need to know where your case study city is located. For example, Mumbai is located on the west coast of India, around a deep bay that has made it an important port for centuries.

Importance of your city

You need to know what factors have made your city important regionally, nationally and internationally.

- Is it the centre of government?
- Is it an important industrial centre?
- Are there historical reasons for its importance?

For example, Mumbai has a long history as a key port, regionally and nationally. Once the centre of India's textile trade, it is now an international leader in IT, finance and media.

Population growth

For your case study city, you should know some facts about its population size and rate of growth. You also need to know about the reasons for population growth, especially natural increase and migration.

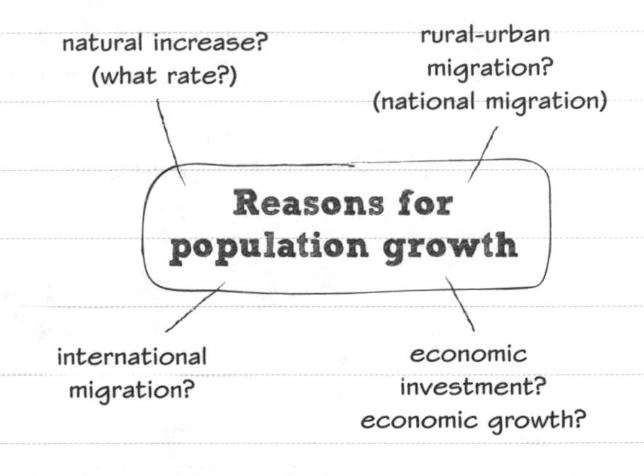

Mumbai's population growth

- Estimated population in 2015: 13 million
- Population in 1991: 9.9 million
- Population growth rate: 2.9 per cent per year
- Migration – 1000 national migrants (from elsewhere in India) arrive each day
- Migration: 90 per cent of migrants are coming from rural areas in India
- Over the last 20 years, migration has contributed to 85 per cent of Mumbai's growth
- Natural increase is now more important as a cause of population growth in older, congested parts of the city

Worked example

Study the table, which shows the results of a survey of recent migrants to Mumbai, asking them why they had moved to the city. Suggest a definition for urban–urban migration. **(1 mark)**

	Rural–urban (%)	Urban–urban (%)
To find a job	68.4	47.2
To start a business	5.5	7.6
Education	2.8	4.9
Marriage	8.0	17.4
Job transfer	3.4	6.9

People moving to Mumbai from another city in India.

Now try this

Put together a fact file about population in your case study city like the one shown for Mumbai.

Non-UK city: opportunities

For your case study of a major city in an LIC or NEE, you need to know how urban growth has created social opportunities and economic opportunities.

Social opportunities

Compared to rural areas, growing cities offer much better social opportunities.

- Health care is available when people need it, even if it costs more than they can afford.
- Primary education is often free for children in LICs and NEEs.
- Piped water and electricity are provided, for a fee, to many residents (though not all).

Economic development

Economic opportunities in growing cities attract migrants from rural areas. This creates a **multiplier effect** for economic development.

Multiplier effect – when an increase in one economic activity causes an increase in others.

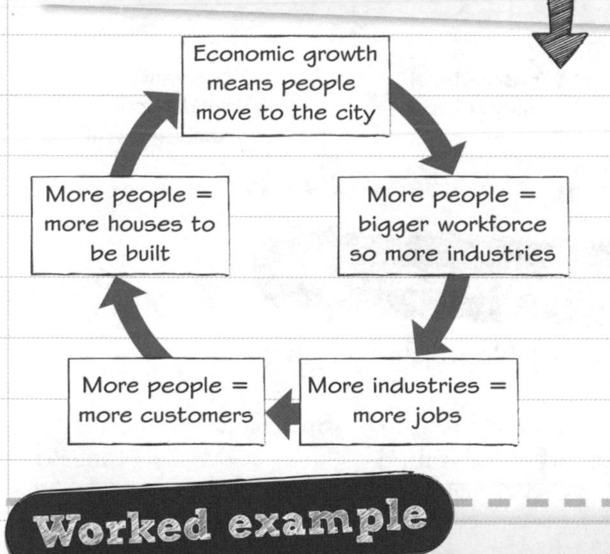

Economic opportunities

Most people move from rural areas to growing cities because of jobs.

- Some LIC and NEE cities grow because they are centres for industries like manufacturing. As more people migrate to the cities, industries grow and need more workers.
- In other growing cities, the service sector is the biggest economic sector. With millions of people in the city, there is huge demand for services. Many service jobs are in the **informal employment** sector (where jobs are not officially recognised or controlled).
- As cities grow, there is a huge demand for workers to build offices and apartments.

Economic opportunities

Mumbai has Asia's largest squatter settlement, called Dharavi. More than 1 million people live in Dharavi – an area of 1.5 km^2. There are 5000 small-scale businesses in Dharavi, and 15 000 single-room 'factories'. Almost all are in the informal sector. It has been estimated that Dharavi's businesses generate the equivalent of £350 million for the Indian economy each year.

What example shows the economic opportunities in your case study city?

Worked example

This photo shows a Mumbai resident working in a pottery, typical of thousands of single-room factories in Dharavi. Suggest **two** advantages of informal sector employment for people who have recently migrated to a growing city in an LIC or NEE. **(2 marks)**

It is easy to set up a business in the informal sector and people keep all the money they earn.

Family members can all work together in the same business – making pots, for example.

Now try this

What example(s) would you use to show the different social opportunities in your case study city?

Non-UK city: challenges 1

For your case study, you need to know about the challenges of slums and squatter settlements, and the challenges of providing clean water, sanitation and energy.

Housing challenges

Rapid population growth means there are far more people arriving in the city than there are affordable houses.

This shortage of housing forces people to live in:

- slum housing – often with many people sharing each room
- squatter settlements where people build housing out of any materials they can find.

Where are slums and squatter settlements in your case study city? What are the challenges of living there?

Slums and squatter settlements

Squatter settlements often develop on land unsuitable for building, such as steep hills or by rivers at risk of flooding.

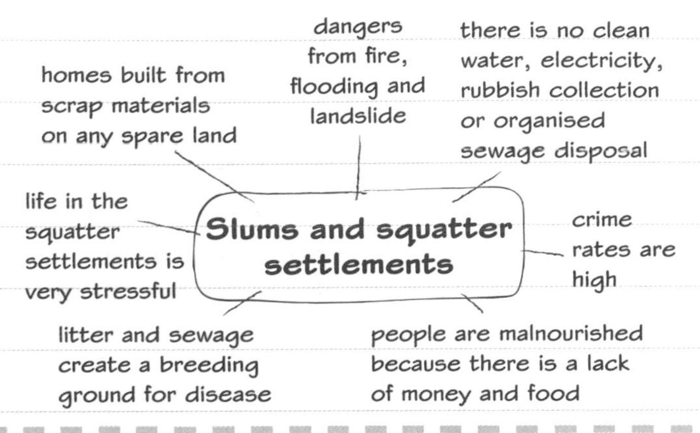

Water and sanitation challenges

Squatter settlements often do not have a piped water supply or a piped sanitation service.

- People can buy bottled water to drink, but it is expensive.
- Some people take water from nearby rivers or streams, which are often polluted and carry diseases.
- People often go to the toilet in waste ground. These waste areas smell bad and diseases can spread.

Energy supply challenges

Growing cities can have problems with energy supply.

- Those connected to the electricity grid can experience frequent power cuts because of the high demand for energy.
- Squatter settlements are often not connected to the grid and use illegal connections instead.
- Many people in LICs buy charcoal for cooking, which can add to city air pollution.

Worked example

This photo shows an area of squatter settlement being cleared in Mumbai. Suggest **one** advantage and **one** disadvantage of this method of managing rapid urban growth. **(2 marks)**

One advantage of clearing squatter settlements is that residents can be rehoused in better quality apartment blocks.

One disadvantage of clearing squatter settlements is that residents often do not want to move to new areas that are far away from where they work.

Now try this

What examples would you use from your case study city for **(a)** challenges of slums and squatter settlements and **(b)** challenges of providing clean water, sanitation systems and energy?

Non-UK city: challenges 2

For your case study, you need to know about the challenges of providing access to services, reducing unemployment and crime, and managing environmental issues.

Providing access to services

Health care

- Poorer residents often cannot afford health services.
- Some city governments send health workers into slums and squatter settlements to provide free care.

Education

- Families sometimes prefer or need children to work rather than go to school.
- Some city governments give families grants so they can afford to send their children to school; charities set up schools in squatter settlements.

Unemployment and crime

Some major NEE cities are now services-based. **Formal sector** jobs need skills most residents do not have, increasing the **informal sector**.

The informal sector has disadvantages:

- low pay that is not always regular or reliable
- dangerous working conditions
- no benefits or security: if people get ill and cannot work, they have no money
- children may be put to work instead of going to school
- poverty makes crime difficult to avoid: gangs offer young people money and security.

Environmental issues

Congestion in Mumbai

Congestion – huge traffic jams can damage the economy, and create harmful air pollution.

Air pollution – high volumes of traffic, heavy industry and power stations create major problems with air pollution in some cities.

Waste – poorer areas having no access to waste disposal, or people are forced to live and work where the rest of the city dumps its waste.

Water pollution – people in poorer areas are forced to use nearby water sources as toilets. Controls on industrial pollution may not be effective.

Reducing unemployment

City governments would like more people in formal sector jobs that pay tax – but it is hard to achieve while skills are low. Governments promote education to teach the skills needed.

Reducing crime

Slum or squatter settlement areas are often taken over by criminal gangs and can turn into no-go areas for city police. City governments have to set up heavily armed police units to invade these areas and suppress the gangs.

Worked example

Describe **one** way that environmental issues are managed in a city you have studied. **(4 marks)**

SPARC is an Indian non-government organisation (NGO) that works in Mumbai slums to build new toilet blocks, connected to city sewers and water supplies. The local community helps to construct the toilet block. The SPARC toilet blocks have electric lights, making them safer to use at night, and separate toilets for children. In the past 5 years, SPARC has provided 800 community toilet blocks, each containing 8 toilets.

Now try this

What example(s) from your case study city would you use to show **(a)** what the environmental challenges are and **(b)** the extent to which these are being managed?

Planning for the urban poor

You need to know an **example** of how urban planning is improving the quality of life for the urban poor. This can be from your LIC/NEE city case study (but does not have to be).

Improving housing

Urban planning can improve quality of life for the urban poor by improving their housing. Ways to do this include:

- upgrading squatter settlements with proper building materials
- clearing squatter settlements and rehousing residents in new blocks
- giving squatter-settlement residents legal ownership of their land and help in improving their homes
- providing electricity, sanitation and water to squatter settlements.

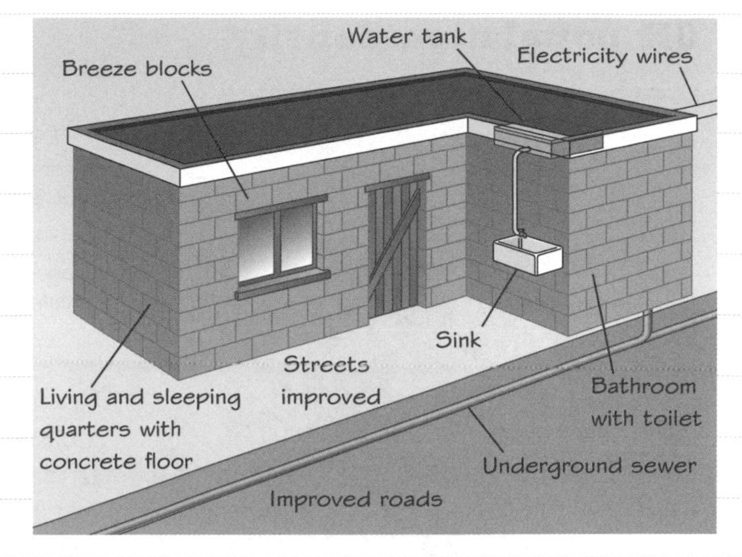

Types of urban planning

Top-down strategies are large-scale government schemes. Bottom-up strategies are small and community-based.

Example

Top-down

The Favela Bairro Project in Rio de Janeiro had a budget of $1 billion and improved living conditions for 250 000 people in 73 squatter settlements, rebuilding shacks into brick buildings, running training programmes and providing waste disposal services for squatter settlements.

Example

Bottom-up

In India, the Self-employed Women's Association runs the all-women SEWA Bank. After a year of saving with the bank, women can apply for a loan. House-rebuilding loans are used to upgrade homes (including 'monsoon proofing' roofs), connecting to piped water or electricity infrastructure, building toilets and providing concrete floors.

Worked example

Using an example, explain **one** way urban planning can improve quality of life for the urban poor. **(2 marks)**

One problem for many poorer people living in LIC and NEE cities is flooding, because they live in areas that are at high risk of flooding. An urban-planning solution is to give building companies tax breaks if they include lakes and green spaces in their housing development to help reduce runoff and absorb floodwater.

Now try this

Using an example, discuss how far urban planning in developing countries (LICs, NEEs) has improved the quality of life for the people who live there. **(6 marks)**

Urban UK

81.5 per cent of the UK's population lives in urban areas: 45.7 million people. You need to know about the general distribution of population in the UK, and what its major cities are.

UK population density

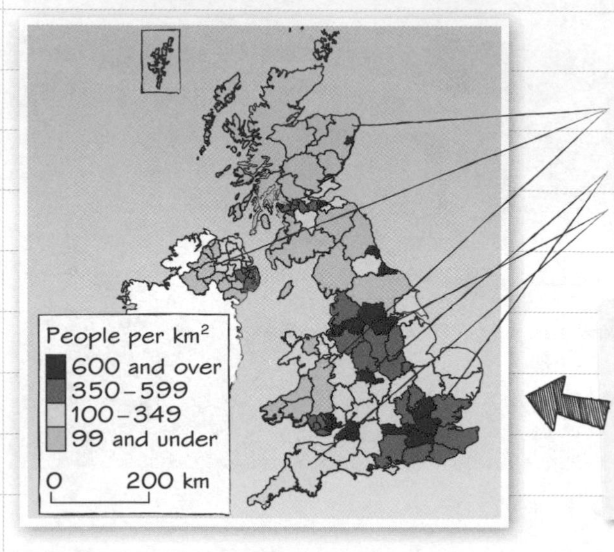

The population distribution of the UK is **uneven**.

Lower population densities mostly in Wales, northern England, Scotland and western Northern Ireland

The West Midlands and south-east England are densely populated

A few other areas of dense population, for example in south-west England and north-east England

Reasons for higher population density include physical factors (such as flatter land, warmer winters) and human factors (such as employment opportunities, trade links and accessibility – transport links).

UK urban areas

There are five urban areas that have more than 500 000 residents: London, Birmingham, Leeds, Glasgow and Sheffield. The next five biggest are Bradford, Liverpool, Edinburgh, Manchester and Bristol, which all have populations over 400,000.

London is by far the biggest city by population in the UK. It has over 7 million people. Birmingham, the next biggest, has 1 million. Birmingham is currently the UK's only other city with a population of 1 million or more.

Main urban areas in the UK

Worked example

Explain **two** reasons why some areas of the UK are densely populated. **(4 marks)**

South-east Wales is densely populated due mainly to the historic development of industries such as iron and steel there, plus the location of major ports such as Cardiff.

South Wales today is easily accessible by motorway (M4), railway and by air, which makes it a good location for businesses.

Now try this

Study the population density map of the UK above. Describe the distribution of the UK population. **(2 marks)**

UK city: location and migration

You need to know a **case study** of a major UK city to show how urban change leads to different opportunities and challenges. Here we look at Birmingham.
Revise the city that you studied in class, using the prompts here to help you.

City in context

Make sure you know about why your case study city is important:

- in the UK
- in its region
- globally.

For example, Birmingham is:
- located in the centre of **England**, with fast motorway and rail connections
- the main city of the West Midlands **region**, a major centre of manufacturing
- home to people from all over the **world** – one third of residents are from minority ethnic backgrounds.

People travel to Birmingham's Bullring Shopping Centre from all over the UK. Located in Birmingham's CBD, the Bullring was redeveloped in 2003 at a cost of £500 million. Around 38 million visitors come to the Bullring each year.

Migration impacts

You need to know what impact national and international migration have had on the growth and character of your city.

- Population growth means more housing, services and jobs are needed.
- National migration – within the country – includes students moving to university cities.
- Most international migrants to the UK move to cities. International migration significantly boosts city economies.
- Cities attract skilled migrants because of their range of job opportunities. Unskilled migrants often live in deprived city areas: cheaper housing.

Migrants are usually young: changing cities' age profiles	Migrants help fill skills shortages: boosting the economy
Migrant families often have children: helping cities grow	Highly motivated, hard-working migrants boost productivity
More children in a city puts pressure on education services	Non-migrants can feel excluded: impacts on city voting patterns

Birmingham and migration

Around 42 per cent of Birmingham's population is from an ethnic minority population, many from Pakistan and India. New arrivals can feel more at home among communities of people from their old country.

- Around 40 per cent of Birmingham's residents live in areas described as among the most deprived 10 per cent in England.
- The areas of deprivation are found in a ring around the city centre.

Worked example

Identify **two** ways in which recent immigrants to a city impact on the character of city areas. **(2 marks)**

Immigrants can bring new cuisines with them and that can mean new types of restaurants and new types of food in shops and supermarkets.

New immigrants may strengthen the small business sector by increasing the number of shops and widening the types of services provided in the area.

This answer recognises that 'character' refers to what features make a city stand out, what makes it distinctive.

Now try this

Explain the impact that migration has had on the growth of your case study city. **(2 marks)**

Had a look ☐ Nearly there ☐ Nailed it! ☐

UK city: opportunities

For your case study, you need to know about how urban change has created social, economic and environmental opportunities: urban greening.

Social opportunities

How has urban change in your case study city led to new social opportunities?

Growing cities with ...

a large population

- wide range of recreational and entertainment opportunities
- people can combine shopping, meals out, visit to cinema, theatre, museum or bar

a young population

- strong music scene
- lots of clubs and bars

growth from international migration

- rich cultural mix
- different types of restaurants
- varied shopping opportunities

Birmingham has a large student population (65 000), so it is affected by **studentification.** This can mean social and cultural changes as areas of the city become dominated by students.

Economic opportunities

- In the second half of the 20th century, most UK cities declined as industries moved out and city ports and docks declined: **deindustrialisation.**
- Manufacturing has declined in cities, but financial and business services have expanded in city centres, bringing employment. Globalisation means financial transnational corporations **(TNCs)** locate in major UK cities.
- City governments have invested heavily in creating retail 'experiences' so people come to cities to shop rather than shop online. This creates many opportunities in retail employment in shops, restaurants and cinemas.

How has urban change in your case study city led to new economic opportunities?

Transport solutions

As the population of cities increases and road congestion causes problems, city governments may invest in efficient **integrated transport systems** that bring together railways, buses and trams to make travel across the city simpler and cheaper. Employers can find workers from across the city and there are benefits from reduced pollution.

Environmental opportunities

Urban greening means increasing the number of plants in urban areas. Urban greening is a cheap and effective way for cities to improve the urban environment.

Three benefits of urban greening include: shade during hot weather; reduction of surface runoff and urban flooding; improving air quality by absorbing air pollutants.

Worked example

Outline **one** way that urban greening provides opportunities for residents in a UK city you have studied. **(2 marks)**

Birmingham's citizens, especially in inner-city areas, have worse-than-average health for the UK. Birmingham Active Neighbourhoods is a project involving four large housing estates. The aim is to increase residents' use of Birmingham's green spaces in order to improve their health.

Now try this

Discuss the social and economic opportunities in your chosen city in the UK. **(6 marks)**

UK city: challenges 1

For your case study, you need to know about how urban change has created social, economic and environmental challenges.

Economic challenges: deprivation

Urban deprivation means some people in cities not having access to the same resources and opportunities as others. Urban deprivation is an economic challenge as well as a social one because people in deprived areas often do not have qualifications needed for jobs.

Measuring deprivation

The Index of Multiple Deprivation (**IMD**) scores small areas across the whole of the UK on a different range of measures. All the areas are ranked from 1 (most deprived) to 32 482 (least deprived).

Birmingham's inner-city wards are among the most deprived in any city in England. What is deprivation like in your case study city?

Why do cities have deprivation 'hot spots'? Reasons include:

- a lack of jobs as industry moves out of the inner city, leaving behind those residents who cannot afford to move
- the inner city areas have old housing, which residents cannot afford to maintain: lowering environmental quality
- crime increases with deprivation, reducing investment in the inner city.

Social challenges: inequalities

City residents do not all have access to the same opportunities.

- **Housing** – in wealthy areas of the city, people can get mortgages for high-quality homes and afford a high quality of life. In deprived areas, people often rent lower-quality housing from the council or private landlords. People may not be able to afford a reasonable quality of life.
- **Education and health** – deprived areas have a lower percentage of young people achieving qualifications at school. People from wealthier areas live longer and have fewer lifestyle-linked diseases.
- **Employment** – deindustrialisation means fewer jobs are available if people have few qualifications.

What are the social and economic challenges in your case study city?

Environmental challenges: building

Urban **dereliction** (abandoned and run-down buildings and infrastructure) can be tackled by pulling down the buildings and redeveloping the **brownfield** site. Inner-city brownfield sites are often valuable because they are close to the city centre, but the sites are often contaminated and are very expensive to clear and develop.

How has dereliction been tackled in your case study city? Have brownfield sites been developed? Or do developers prefer cheaper **greenfield** sites on the rural–urban fringe?

Worked example

Describe **one** way that waste disposal leads to environmental challenges in a UK city you have studied. **(2 marks)**

In Birmingham, the rapid growth of the city (2.5 per cent in 2012–13) creates a lot of rubbish. Birmingham City Council wants to increase the amount of recycling because 70 per cent of waste currently has to be buried in landfill sites.

Now try this

Explain how economic change is involved in creating inequalities in your case study city. **(4 marks)**

Had a look ☐ Nearly there ☐ Nailed it! ☐

UK city: challenges 2

For your case study, you need to know about the impact of urban sprawl on the rural–urban fringe, and the growth of commuter settlements.

Urban sprawl

The rural–urban fringe is under pressure from urban sprawl.

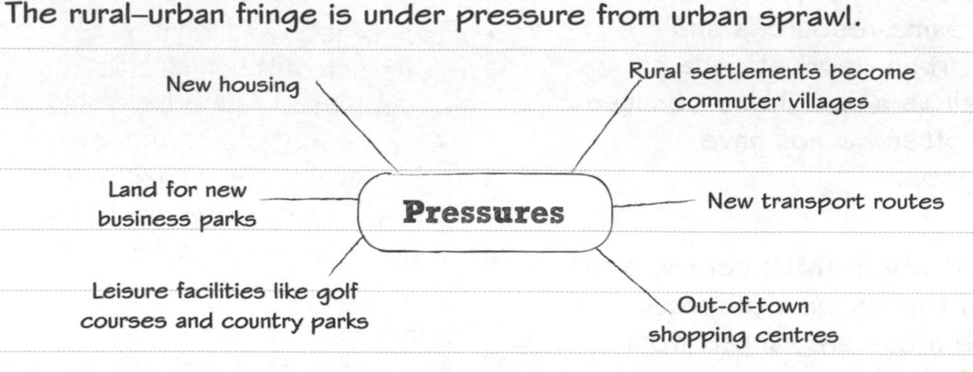

The **rural–urban fringe** is an area around a city or town where urban land use mixes with rural land use and competes with it. **Urban sprawl** is the spread of urban areas into the surrounding countryside.

Reasons for development

- Developers prefer rural areas on the edge of towns because there is more space to expand. Rural land is often cheaper.
- People want to live in a pleasant rural environment and commute into work in the city.
- Out-of-town shopping centres utilise cheaper land and locate near major roads to attract customers from all over the region.

Impacts of development

- Traffic noise and development disturbs wildlife and changes the character of rural areas.
- Traffic congestion increases as commuters travel into town.
- The social and economic character of villages changes when they become commuter villages.
- Agricultural land and green areas are lost.

Commuter villages

Commuter villages are close enough to a town or city for people to travel to the city to work each day.

Worked example

Study the OS 1:25 000 map extract, which includes the Birmingham Business Park.

Identify **two** opportunities that this city location offers for a business park development. **(2 marks)**

This location on the urban–rural fringe has excellent transport links (a motorway and major roads) and plenty of flat land for business park expansion.

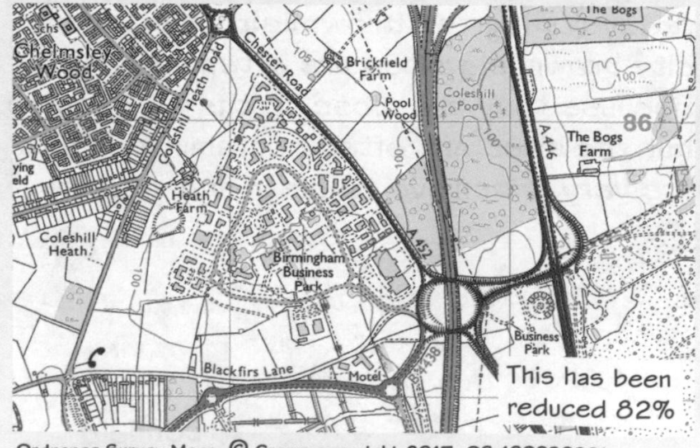

Now try this

Outline the challenges created by urban sprawl in your case study city. **(3 marks)**

UK urban regeneration

Urban regeneration projects aim to create new opportunities in urban areas that are experiencing economic and social problems. You need to know an **example** of an urban regeneration project in the UK, from either your case study city or somewhere else in the UK.

Example: Urban regeneration in Sandwell, in the West Midlands

In the 1990s, many of Sandwell's manufacturing industries closed, leading to unemployment, deprivation, inequality and environmental challenges.

This example shows the sort of information you need. Revise this if you do not have enough detail in your notes for the example you did in class.

Reasons for regeneration

- large areas of ground were poisoned by mercury and cadmium
- the air was badly polluted
- there was very little green space
- over 23 per cent of council housing was unfit for habitation
- In 2010, Sandwell was the twelfth most deprived area in the UK.

Main features of the project

- the polluted land has been cleaned up
- creation of urban reserves, e.g. RSPB Sandwell
- some reclaimed land used for new industries, such as automotive parts
- 8500 new jobs; 45 km of new roads; 300 new industrial units
- new schools have been built or refurbished
- the new Midland Metro Tramway increases access to Birmingham and Wolverhampton.

Who helped improve Sandwell?

The Black Country Development Corporation and Tipton Challenge (paid for by local and central government) and 21 Urban Regeneration Companies reclaimed the land and the New Deal for Communities programme helped improve housing and facilities.

Worked example

Outline the main features of an urban regeneration project that you have studied. **(4 marks)**

At Longbridge in Birmingham, 2000 new homes have been built on a brownfield site to help meet local demand for new homes. Bournville College has also relocated to the site with a new £66 million facility, and new parks are being created for urban greening. Most importantly, 10 000 new jobs will eventually be created as a result of the regeneration project. There is a new retail centre, including a large Sainsbury's supermarket and community facilities.

Now try this

This photo features a £3 billion regeneration project redeveloping London's Elephant and Castle area. Suggest **one** advantage and **one** disadvantage of urban regeneration for local residents. **(2 marks)**

Sustainable urban living

Sustainable urban living means being able to live in cities in ways that do not pollute the environment and using resources in ways that ensure future generations also can use them.

More sustainable urban living

Ways to make urban living in the UK more sustainable

Use brownfield sites for development to:
- improve appearance of these areas
- create new green spaces in the city.

Reduce waste by:
- recycling – 90% of household waste is recyclable
- reusing, e.g. bottles, plastic bags, etc.

Renovate old buildings to:
- enhance the appeal of the area
- improve energy efficiency
- reduce wasteful use of water.

New housing that is:
- affordable to rent or buy
- energy efficient
- efficient in water use and storage.

Improve public transport systems by:
- linking bus, tram and rail routes
- providing feeder services to housing estates
- using low carbon emission vehicles.

Involve communities in local decision-making:
- consult local people instead of imposing plans
- put people first; ask for and act on their ideas
- foster the growth of a community spirit.

Creating green space

Creating more green spaces in cities helps increase sustainable urban living in several key ways.

- Green spaces encourage people to walk in cities, and take other forms of exercise, too. This helps people's physical health and mental health.
- People are more likely to walk or cycle to work if their journey is through green spaces rather than along busy roads. This reduces traffic congestion and traffic emissions.

All the concrete, glass and tarmac in cities heats up to create an **urban heat island**, raising city temperatures. Green spaces reduce this effect (and also absorb CO_2).

This housing development in Salford, Manchester, has been made more energy-efficient with high levels of insulation and solar power panels

Worked example

Explain **two** ways in which urban living in the UK can be made more sustainable. **(4 marks)**

Cities that waste water or pollute water sources are not sustainable. City water use can be made more sustainable by increasing the use of water-efficient technology in homes and businesses. For example, water that has been used for drinking and washing can be recycled and used again (for cooling in city air conditioning or energy production, for example). Cities produce a lot of waste. Using landfill sites to dispose of this waste is not sustainable: there are environmental impacts and new landfill sites have to be found. Urban recycling schemes are a sustainable solution. These include community-based recycling centres (for example, located next to supermarkets) and recycling bins with weekly or fortnightly collections from the kerbside. UK urban areas recycle 40 per cent of their waste; there are targets to increase this.

Now try this

Study the photo of the housing development. Identify **two** ways in which urban housing can help make cities more sustainable. **(2 marks)**

Urban transport strategies

You need to know how urban transport strategies are used to reduce traffic congestion.

Congestion solutions

Too many cars on the roads leads to congestion and pollution. Ways to reduce congestion and pollution would be to:

- widen roads to allow more traffic to flow more easily
- build ring roads and bypasses to keep through traffic out of city centres
- introduce a congestion charge
- introduce park and ride schemes
- encourage car-sharing schemes
- have better public transport, cycle lanes, cycle hire schemes, etc.

Park and ride schemes reduce congestion

Worked example

Study the graph. Part of London's Transport Strategy is to increase cycling so that 5 per cent of all journeys in London by 2026 are made by bicycle.

Projected cycle journeys
Actual cycle journeys
Million cycle journeys: 0, 0.2, 0.4, 0.6, 0.8, 1.0, 1.2, 1.4, 1.6
Year: 1995, 1997, 1999, 2001, 2003, 2005, 2007, 2009, 2011, 2013, 2015, 2017, 2019, 2021, 2023, 2025

(a) Suggest **one** advantage for London of increasing the number of journeys made by bicycle. **(2 marks)**

If people use bikes to get to work or school in the mornings, it will reduce the number of motor vehicles on the road, reducing congestion.

(b) Suggest **one** way in which people could be encouraged to cycle instead of use cars to make journeys in London. **(2 marks)**

More cycle lanes that were separated from traffic would make people feel safer when they cycle and encourage more people to cycle.

The London congestion charge

This is an £11.50 daily charge for driving inside a specific zone between 7am and 6 pm, Monday to Friday. Automatic number plate recognition systems check the vehicles travelling inside the zone and send a bill to anyone who has not already pre-paid the charge.

The congestion charge raised £2.6 billion in its first 10 years of operation, half of which was used to improve public transport and cycle schemes in London. Traffic was reduced by 10 per cent, but traffic speeds have still continued to fall due to traffic jams in the congestion zone.

Now try this

1. Read the information about London's congestion charge. Identify **two** ways in which the charge aims to reduce congestion. **(2 marks)**
2. Discuss the effects of urban transport strategies on a major UK city you have studied. **(6 marks)**

Had a look ☐ Nearly there ☐ Nailed it! ☐

Measuring development 1

There are lots of different ways to measure development. Here are some examples:

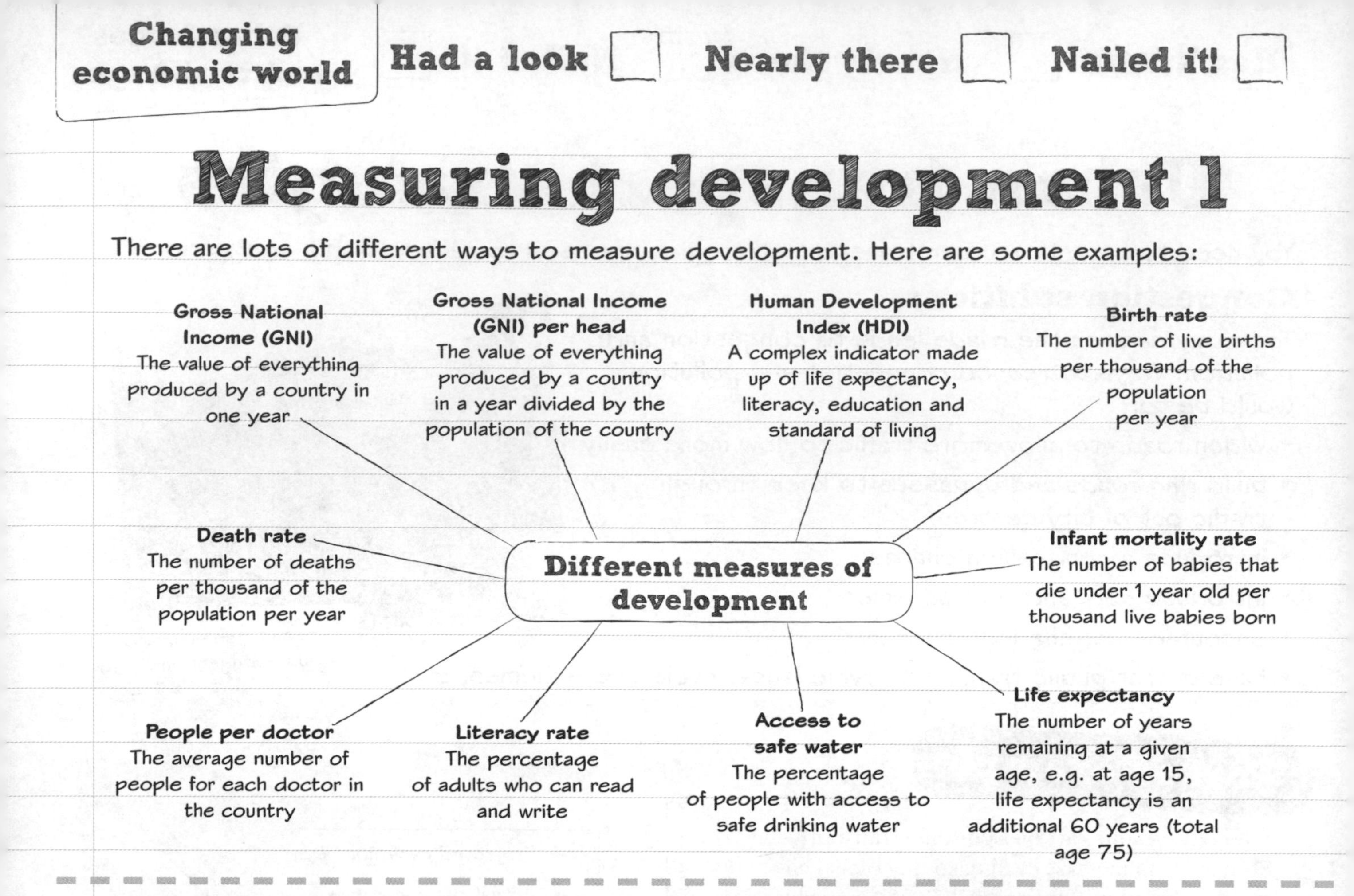

Limitations

GNI gives an average picture across a country, so if there is a rich elite and a lot of very poor people, this would not show up.

The governments of some countries may not collect GNI data accurately, which can make data unreliable. Data for GNI can also become out of date quickly if countries experience big changes: for example, large-scale migrations of people.

HDI scores for different countries are often displayed on a **scattergraph**.

For more on scattergraphs, see page 138.

Worked example

Study this table which shows HDI ranks and figures for four countries.

Rank	Country	HDI 2014
1	Norway	0.944
50	Russian Federation	0.798
100	Tonga	0.717
150	Swaziland	0.532

What does HDI measure? **(2 marks)**

HDI measures human development and ranks countries into order according to their HDI score. It combines statistics on life expectancy, literacy, education and standards of living for each country.

Now try this

Study the table on this page showing HDI scores for different countries. Explain whether you would expect the GNI per head for each country to correlate with the HDI scores. **(2 marks)**

Measuring development 2

There are different ways of classifying parts of the world according to their level of economic development and their quality of life.

The North–South divide

This is a way of talking about differences in development between the global North – North America, Europe, Russia – and the economically poorer countries of the South.

One problem with this classification is that it is general. There are many differences in development within both South and North.

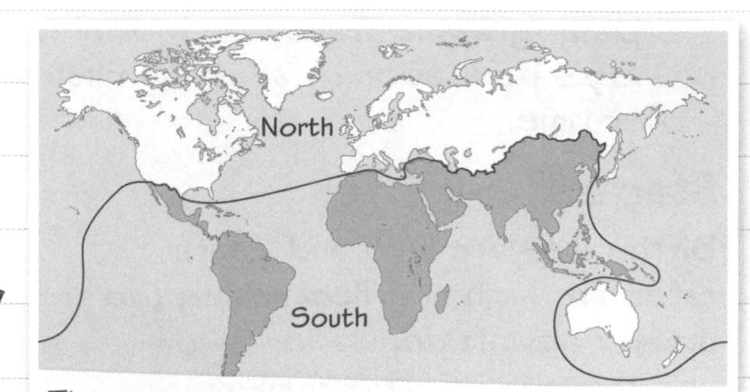

The North–South classification of parts of the world, now seen as outdated

HICs, NEEs and LICs

The World Bank categories for development give higher income countries (HICs) a GNI per head of US$12 476 or more. Lower income countries (LICs) have a GNI per head of US$1025 or less. The World Bank uses two other categories for the 'middle income' newly emerging economies (NEEs).

Some countries in the middle income category are NEEs but others are not developing quickly. The World Bank categories are updated yearly.

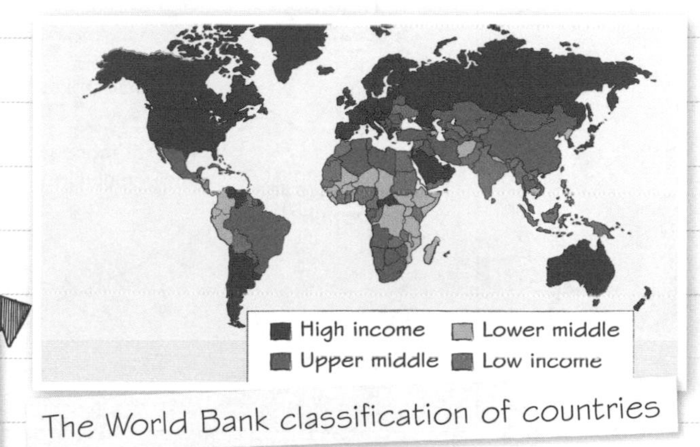

The World Bank classification of countries

Human Development Index (HDI)

More detailed classifications, such as an HDI map, are not as simple to use as a North–South divide or an HIC/NEE/LIC classification, but they do capture more of the differences within categories. They recognise social variations rather than being wholly based on a county's economic performance.

Standard of living and quality of life

- Standard of living is an economic measure. Do people have enough money to live on?
- Quality of life is a social measure. Do people have a long and healthy life?

Worked example

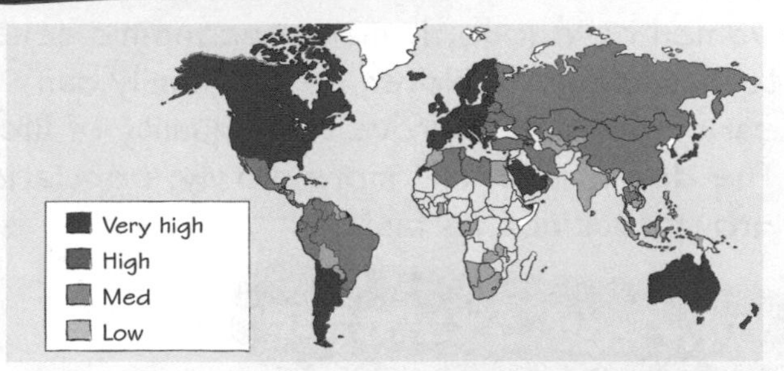

This map classifies countries into very high, high, medium and low human development.
Identify **one** NEE with high human development and **one** with medium human development. **(2 marks)**

High: Brazil   Medium: India

Now try this

Compare the World Bank map with the human development map on this page. Suggest **one** reason why countries that, economically, are LICs can sometimes have high human development. **(2 marks)**

The Demographic Transition Model

The **Demographic Transition Model (DTM)** describes a theory of how development affects a country's population by affecting birth rates and death rates – and therefore total population – over time.

Stage 1

Birth rates are high and death rates are high and **fluctuating** (go up and down): population is low and there is no development.

Stage 2

Death rates start to fall due to developments in health care and sanitation. Birth rates stay high because people still have insecure lives and having many children helps a family survive. This generates rapid population growth.

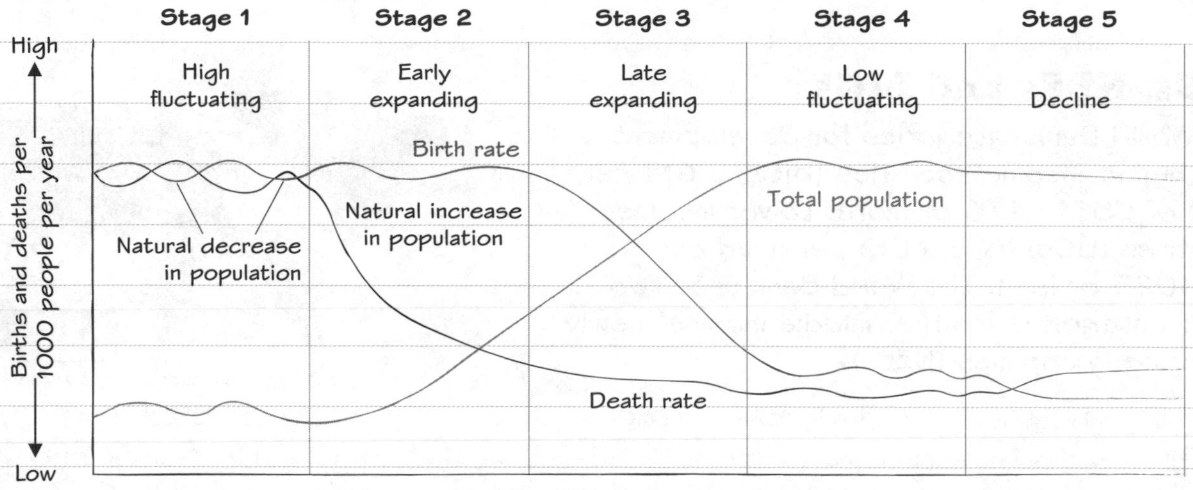

Stage 1 – High fluctuating
Stage 2 – Early expanding
Stage 3 – Late expanding
Stage 4 – Low fluctuating
Stage 5 – Decline

High
Births and deaths per 1000 people per year
Low

Birth rate
Natural decrease in population
Natural increase in population
Total population
Death rate

Stage 3

Economic development, improved education and the availability of contraception means families decide to have fewer children. When women have jobs, it makes economic sense to have fewer children so the family can earn more and improve their quality of life. The death rate falls more slowly; population grows, but not as fast.

Stage 4

(UK is at this stage)

Low birth rates and death rates. Large population but growth slows. Births and deaths are close in numbers, but fluctuate.

Stage 5

Birth rates fall below death rates and population starts to decline.

Worked example

Suggest **one** reason why increasing economic development means families decide to have fewer children. **(2 marks)**

Children are often economic assets in countries with low economic development. However, when people live in NEEs like India, the cost of raising and educating a child rises. Families may then decide to have fewer children so they can maintain higher standards of living and quality of life.

Now try this

Replacement level fertility is the fertility level at which the number of births matches the number of deaths. In which stage of the DTM would you expect to find replacement level fertility? **(1 mark)**

Uneven development: causes

The DTM sets out a model of development for all countries to follow; in reality development is uneven around the world. There are physical, economic and historical causes for this.

Factors leading to uneven development

Physical

Countries find it more difficult to develop when they:

- are landlocked and can't benefit from trade by sea
- are blighted by tropical diseases which affect people's ability to work
- have poor soils and arid conditions
- suffer from natural hazards, e.g. earthquakes, and do not have the money to repair all the damage caused.

Economic

Countries find it more difficult to develop when:

- global trade favours already developed countries
- tariffs make trade more expensive
- they produce mainly primary products which don't make much money
- they are in debt and have to spend money on interest payments rather than development.

Historical

Countries find it more difficult to develop when they:

- were colonised by European countries in the 19th and early 20th centuries: this meant their economies were developed just to produce raw materials for manufacturing in European countries
- have long histories of conflict, for example, civil war within the country. Wars mean no stability for economic development, refugee crises and governments spending what money a country has on arms and soldiers.

Worked example

Explain how debt can be an obstacle to development. **(4 marks)**

When poor countries are in debt, they use up money paying the interest on the debt rather than spending it on development projects. Debt can also be an obstacle in everyday life. When farmers have a poor crop they are often forced into debt to survive. Then what money they have goes on paying interest instead of educating their children, for example, or improving their farms.

Now try this

Explain why countries which largely rely on exporting primary products are at a disadvantage in world trade. **(4 marks)**

Things to think about for this question: LICs that rely on primary products are often competing with each other to supply HICs and NEEs with the same product. How would that make it easier for HICs to demand a lower price? HICs and powerful NEEs have set up subsidies to protect their own industries and farmers. How might that disadvantage LICs trying to trade with them?

Uneven development: consequences

Uneven development leads to disparity (differences) between countries and within countries in terms of how much money people have to live on and also in terms of health. One reason for international migration is uneven development, as people move countries in order to access a better quality of life.

Uneven wealth

The diagram below shows how the richest 20 per cent of the world's population have over 80 per cent of global income.

- The richest 20 per cent is mainly the world's HICs, but inequalities exist within HICs too. For example, in the UK the richest 10 per cent earn more than 20 times more a year than the poorest 10 per cent.
- There are inequalities within NEEs and LICs, too. Inequalities within NEEs are often extreme. For example, in 2015 India had a GNI per head of $6030 (the UK was $40 610). Yet India also has the third largest number of billionaires in the world: 111.

Richest

Each band = 20% of the world's population

Poorest

World population	World income
Richest 20%	82.7%
Second 20%	11.7%
Third 20%	2.3%
Fourth 20%	1.9%
Poorest 20%	1.4%

This diagram shows global income divided between five equal shares (**quintiles**) of the world's population. Data is for 2013.

International migration

Uneven development gives people in poorer countries an incentive to migrate to richer countries where they can earn more and enjoy better health care and a higher quality of life (pull factors). The tensions produced by uneven development can also contribute to crises and conflicts (push factors). Refugees escaping conflicts can be a consequence of uneven development.

Worked example

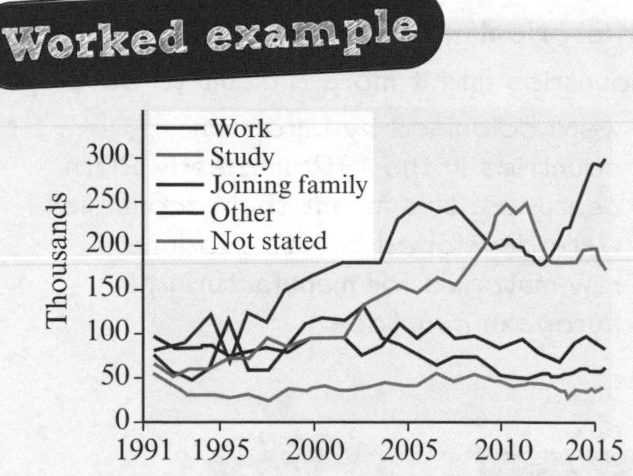

This graph shows reasons given by international migrants for coming to the UK.

Using the graph and your own knowledge, identify **two** reasons why international migrants come to the UK. **(2 marks)**

Reason 1: People come to work in the UK because they can earn more here than in the country they originally come from.

Reason 2: People come to study in the UK because they can learn English here, which can improve their chances of getting a good job.

Now try this

Study the pictogram, which shows the number of women who died giving birth in different world regions, 2015.

State **one** advantage and **one** disadvantage of using a pictogram to present this data. **(2 marks)**

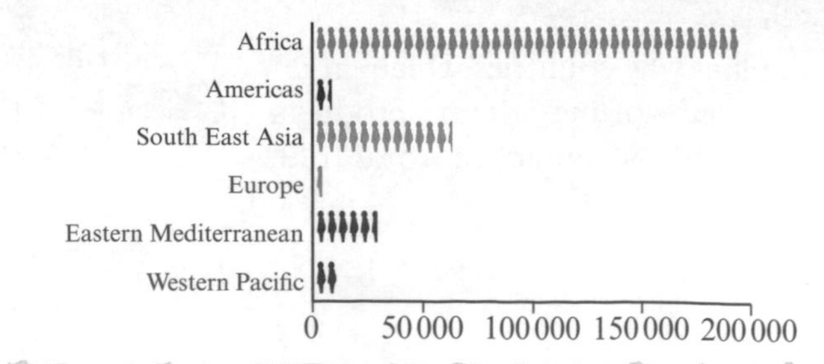

Investment, industry and aid

There are many different strategies for reducing the development gap. These include investment and industrial development aid. Another strategy is tourism. For more on tourism, see page 93.

Top-down development

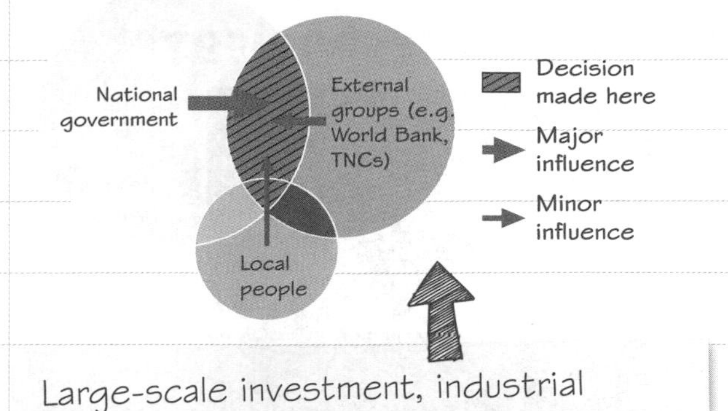

Large-scale investment, industrial development projects, national tourism strategies and international aid are examples of top-down development.

Investment and industrial development

Foreign direct investment (FDI) is when companies in one country invest in companies in another. For example, China is a major investor in Africa: 2000 Chinese companies have invested around $15 billion in African countries, mostly in manufacturing and construction industries and in mining and energy production.

- FDI reduces the development gap because it boosts a country's industrial sector with money, new technologies and new markets.
- However, these investments are designed to benefit the investing company most of all.

Worked example

Study the map, which shows investment (in US billions) by US TNCs in selected African countries.

Explain **one** reason why US TNCs might want to invest in African countries. **(2 marks)**

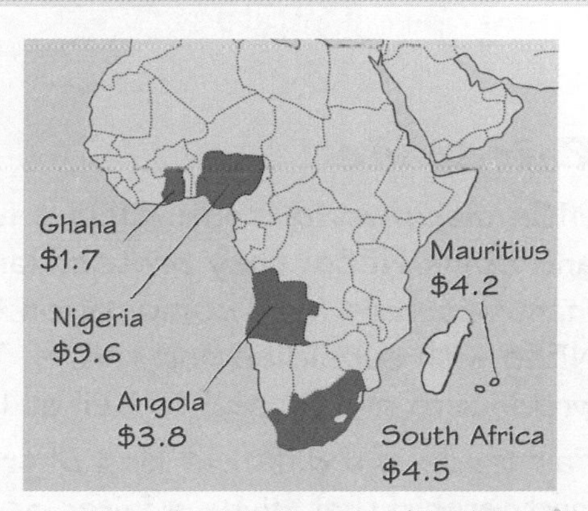

Countries like Nigeria and South Africa have valuable resources – for example, oil and diamonds. US TNCs invest in these countries in order to gain access to these resources – which can also be mined at cheap rates because of lower wages.

Aid

Emergency aid is short-term aid: for example, to help a country recover from natural disaster.

Development aid is used to develop long-term projects, such as immunising children against preventable diseases.

Bilateral aid is when one government sends aid to another government.

Multilateral aid is when governments send aid to international organisations such as the World Bank, which then funds development projects.

Advantages and disadvantages of aid

There are concerns that too much aid money is lost to corrupt LIC governments, and that aid makes LICs dependent on HIC aid, reducing their chances of really developing at all.

However, most aid goes to help the poorest in global society. Large-scale investments in education and health care for poor people have massive benefits for development. This then provides economic benefits for the whole world.

Now try this

In 2015, the UK gave 0.71 per cent of its GNI in international aid. The UK's GNI in 2015 was US$2.552 trillion. How much did the UK give in international aid in 2014?

A US$18.11 billion **B** US$181.19 billion **C** US$181.19 million **(1 mark)**

Technology, trade, relief and loans

The use of intermediate technology, fair trade, debt relief and microfinance loans are all further examples of strategies for reducing the development gap.

Bottom-up development

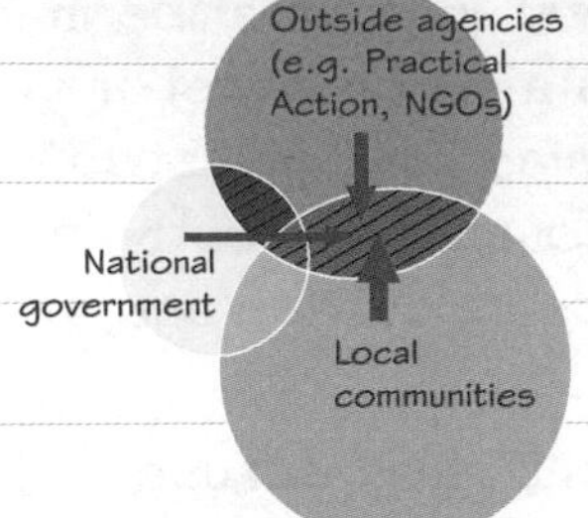

Using intermediate technology, fair trade and microfinance loans are examples of bottom-up development: local communities are in involved in the development strategy.

Intermediate technology

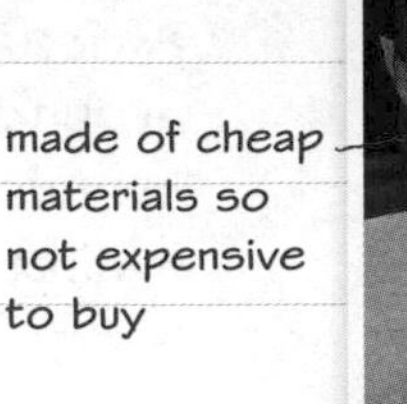

Intermediate technology is suited to the needs, skills, knowledge and wealth of local people. It usually combines simple ideas and materials.

Fair trade

HICs make world trade difficult for LICs and some NEEs: they protect farmers and manufacturers from competition from LICs and NEEs with subsidies and tariffs. Then many LIC producers are forced to sell at low prices.

Fair trade is a different kind of trading partnership that gives a fairer deal to LIC farmers. Customers in HICs pay a **premium** (higher price) for LIC products, knowing this will help improve the quality of life for producers.

Worked example

The photo shows some advantages of using solar cooking technology. Explain **two** ways in which intermediate technology like this can help reduce the development gap. **(2 marks)**

Intermediate technology creates jobs and develops new skills locally, from making and repairing the technology.

Intermediate technology saves local people time and money, so they have more for education and jobs.

Debt relief

The World Bank identifies **Heavily Indebted Poor Countries**. These countries (33 are in sub-Saharan Africa) have unsustainable debts to global banks, which lent them money in the past to fund large-scale development projects. The interest payments on these loans are so big that they keep the countries permanently poor.

Debt relief works by cancelling debts in return for commitments from the debtor countries that they will invest in effective development projects.

Microfinance loans

While foreign direct investment funds large-scale industrial development, microfinance loans are small-scale, designed to help poorer people.

In NEEs and LICs microfinance loans are often important in urban planning projects to help the urban poor.

Look on page 77 at SEWA, a bank that offers microfinance to self-employed women in India to set up businesses or improve their housing.

Now try this

Customers in HICs pay a premium for fair trade products. Explain how fair trade helps reduce the development gap. **(3 marks)**

Tourism

Tourism can be a strategy to reduce the development gap. You need to know an **example** of how the growth of tourism in an LIC or NEE helps to reduce the development gap. **Revise the example you have studied, using the prompts provided here to help you.**

Benefits of tourism

Tourism has major financial benefits for LICs or NEEs that can attract international visitors, especially from HICs.

Top three benefits

1. HIC tourists pay for holidays in foreign currency, which countries need for economic development.
2. Tourists spend money in the local economy, which benefits ordinary people.

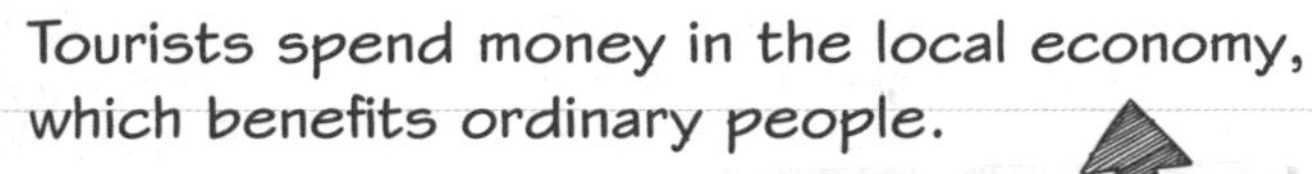

3. Tourism creates many service-sector jobs.

For more on multiplier effects, see page 74.

Tourism can produce a **multiplier effect**.

Reducing the development gap

Money from tourism can be used by the LIC or NEE government to invest in top-down infrastructure projects and industrial development, which help the economy develop.

Foreign currency earned from tourism can be used to pay off international debt; if global banks then decide that the country is a good place to invest, this will encourage FDI. For more on FDI, see page 91.

The jobs created by tourism mean more tax revenues for the LIC or NEE government to spend on improving health care and education.

Create a Fact file for the example you studied.

Worked example

Fact file: Tourism in Kenya

- Kenya has many attractions for international tourists. The top three are safari holidays, coastal beach holidays and cultural tourism.
- Tourism is Kenya's largest foreign currency earner, followed by flowers, tea and coffee.
- Kenya's economic growth relies on tourism. It contributes 10.5 per cent of Kenya's GNI.
- Around 9 per cent of Kenya's working population works in tourism: over 500 000 jobs.

Using the Fact file and your own knowledge, explain why a reduction in tourism would have a negative impact on Kenya's economic development. **(4 marks)**

Kenya depends on tourism as it represents 10.5 per cent of its GNI, employing 500 000 people. A reduction in tourism (for example, after a terrorist attack) would lower foreign currency earnings and tax revenue because there would be fewer tourists in the country and tourist industries would earn less money. This means the government would have less to spend on developing the country economically and socially (for example, health care).

Strategies to reduce negative impacts of mass tourism

Help local people to benefit more from tourism

Tourists encouraged to be more responsible

Tourism companies encouraged to develop less damaging types of activities

Now try this

Study the images above, which show ways the Kenyan tourist industry is tackling disadvantages of tourism. Outline these **three** disadvantages. **(3 marks)**

 Had a look ☐ Nearly there ☐ Nailed it! ☐

LIC or NEE country: location and context

Case study You need to know a **case study** of an LIC or NEE to show the changes that come from rapid economic development. Here we look at India. **You should revise the country that you studied in class, using the prompts here to help you.**

Location and importance

You need to know your country's location and what makes it important, regionally and globally.

For example, India is located in south Asia. It is central to important sea trade routes through the Indian Ocean. It is very large, with over 3 million km^2 of land, and has borders with six countries, including China. With 1.3 billion people it is the second largest country in the world by population. This huge population makes it an important market in its region and the world, and makes its armed forces stronger.

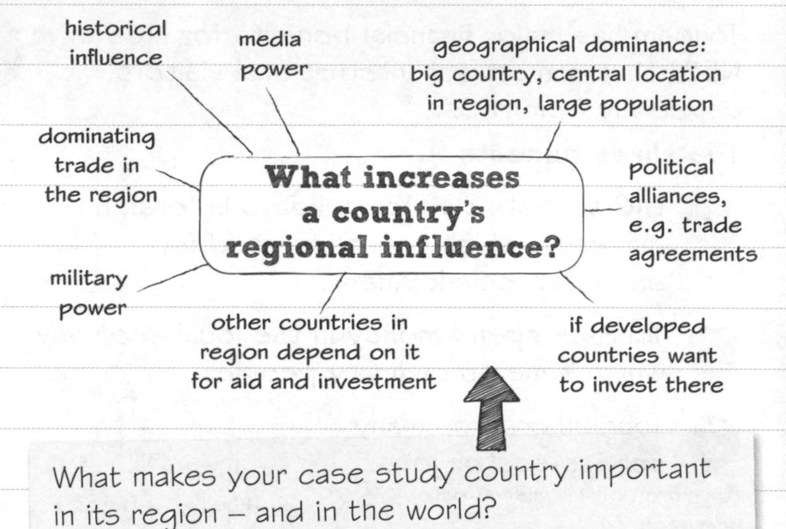

What makes your case study country important in its region – and in the world?

India's rapid economic development includes the world-famous Bollywood cinema industry

Social, cultural and environmental issues

You need to know some key facts about what your country is like: what political system it has, what its culture is like, its key social issues and environmental issues.

The changing industrial structure

You need to know the balance between the different sectors of your case study country's economy – each sector's share of GDP.

For example, India's GDP in 2015 was US$2.095 trillion. The balance of sectors was:

- agriculture: 16.1%
- industry: 29.5%
- services: 54.4%

Worked example

Explain **two** ways growth in manufacturing industry can stimulate a country's economic development.

(4 marks)

When most people farm land to grow food to eat, the country has nothing to export and people are poor. When people work in industry they produce products that can be exported, such as iron and steel. This brings money into the country for development. People in industry get wages, which they spend on more products, creating more jobs (multiplier effect).

Now try this

What is the balance between different sectors of the economy in your case study country? Find out how this has changed over time: for example, since the 1990s.

LIC or NEE country: TNCs

Case study: For your case study, you need to know how **transnational corporations (TNCs)** have been involved in your country's industrial development, and their advantages and disadvantages.

Advantages of TNCs

- TNCs bring investment, modern technology and skills to LICs and NEEs. For example, India got US$30 billion of foreign direct investment in 2012 (China got US$124 billion!)
- TNCs bring big brands into LICs and NEEs, which helps develop a bigger consumer market in the host country. That is important for industrial development as the market buys domestic products, too.
- TNCs often pay more than local companies, which pushes up wages – giving consumers more money to buy things with. Higher wages encourage people to leave farming to work in industry, which means farming becomes more efficient.

Disadvantages of TNCs

- The big brands brought in by TNCs can outsell nationally made products. That can mean job losses in national industries and the growing consumer market may spend money on TNC products – so money leaves the country.
- Investment by TNCs and other foreign investors is not always reliable. For example, India lost US$14 billion of investment after the global crash of 2008, when TNCs cut back on foreign direct investment.
- Lack of regulation of TNC activities can have environmental consequences, like pollution.

Worked example

Study the graph showing the numbers of TNCs operating in India and the amount of foreign direct investment (FDI) in India in 1991–2012. Using the graph, which **two** of the following statements are true? **(2 marks)**

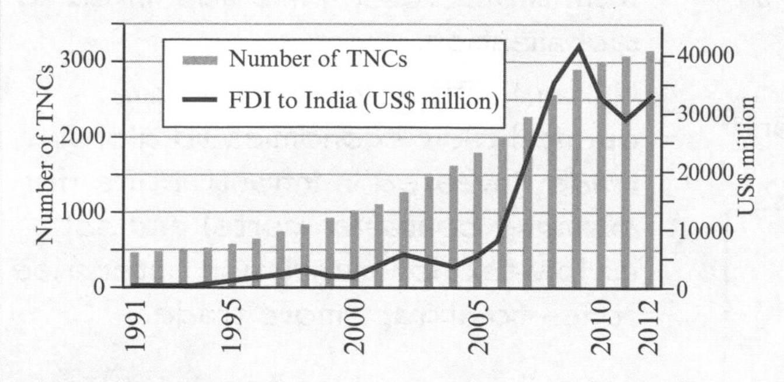

Between 2002 and 2012:

☒ **A** the number of TNCs in India increased by around 1900.

☐ **B** foreign direct investment in India increased by nearly US$6000 million.

☐ **C** the number of TNCs in India increased by around 890.

☒ **D** foreign direct investment in India increased by nearly US$27 000 million.

Questions may test your ability to interpret and extract information from graphs at any point on your exam paper.

In 2011 a major US soft drinks TNC announced an investment of $2 billion in India. What might the impact of this FDI be for Indian soft drinks manufacturers?

Now try this

To what extent have TNCs contributed to the industrial development of a LIC or NEE that you have studied?

(9 marks + 3 SPaG)

 Had a look ☐ Nearly there ☐ Nailed it! ☐

LIC or NEE country: trade, politics and aid

For your case study, you need to know about your country's changing political and trading relationships with the world, and about the impact of international aid on the country.

Political change

Rapid economic development in some NEEs and LICs has changed their political relationships with other countries.

- Countries like India, China and Brazil have become important to the global economy and influential in world politics.
- Growing economic power has increased tensions. Countries like India and China have invested in their military power, which threatens neighbouring countries. India has a military budget of US$32 billion a year.

International aid

Some NEEs have become economically powerful. Countries like China and India give millions of dollars of aid to LICs. But HICs like the USA and UK still send development aid to fast-growing economic giants like India and China. Why?

- Part of the reason is to build good political relationships. **Tied aid** can link aid to trading relationships designed to help HICs.
- The main reason is that there are still huge problems of poverty and inequality in LICs and most NEEs. Aid targets this poverty.

Key terms

Tied aid: aid with conditions, for example, if the UK sends aid to India, India will buy new fighter planes from the UK.

NGO: non-governmental organisation, for example, the charity Action Aid.

For more on types of aid, see page 91.

When cities are growing rapidly property developers often want to evict people living in slums and settlements so they can develop the land. Action Aid helps people to protect their rights to their homes.

Global trade

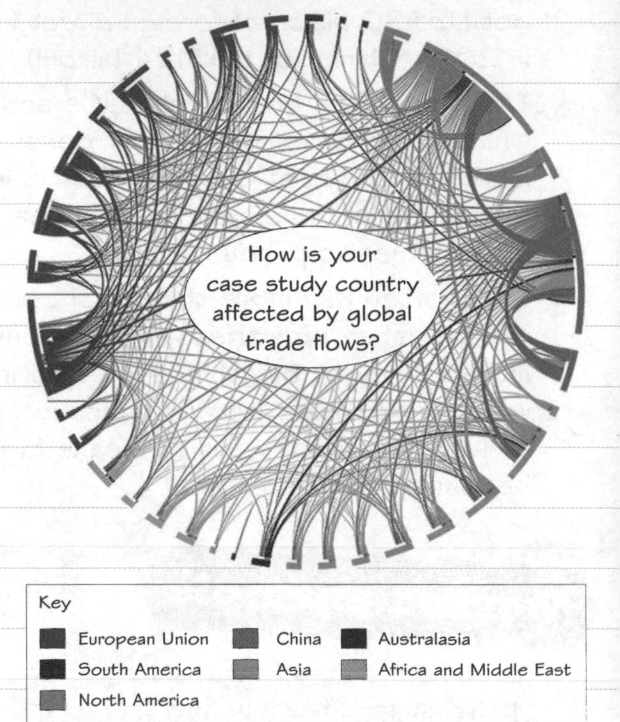

Some LICs and NEEs have benefitted from the increases in trade linked to globalisation.

LIC and NEE governments have opened their economies to global trade, invested in infrastructure (for example, container ports) and set up low-tax, low-regulation enterprise zones to attract more trade.

In 1991, the Indian government cancelled high tariffs on foreign trade. Since 2000, Indian exports to the rest of the world have increased by 620 per cent and imports have increased by 792 per cent. Global trade has been the driver for rapid economic and social change in NEEs like China and India.

Now try this

Find out the top three trading partners of your emerging country – the three countries it exports most to and the three countries it imports most from.

LIC or NEE country: environmental impact

For your case study you need to know about the environmental impacts of economic development and the effects of economic development on people's quality of life.

Environmental impacts

Rapid economic development has damaged the environment.

- Countries need more energy for economic development. More fossil fuels are extracted and burnt, leading to air pollution and a huge increase in carbon emissions.
- Industrial development often involves heavy industries that are polluting.
- Economic development leads to destruction of natural environments, for example deforestation for HEP projects.

Worked example

A World Health Organization report indicates that 13 of the world's most polluted cities are in India. In India air pollution is increasing at 2 per cent per year, causing serious health problems for many citizens.

Study the text about air pollution in India. Suggest **one** reason why rapid economic development increases air pollution. **(1 mark)**

As people get richer they buy cars. Pollution from the cars increases air pollution.

Economic development and quality of life

Rapid economic development has made some people very rich in LICs and NEEs. It has also increased access to education and to health care. However, it has intensified inequalities. In some LICs and NEEs, most people are very poor and their quality of life is very low. You may be able to use your case study of a major LIC or NEE city as examples for this topic, see page 73.

Fact file: Inequality in India

- India's economic growth has created wealth: there are 236 000 millionaires in India.
- However, 300 million Indians live in poverty: defined as living on less than US$1.25 a day.
- The Gini coefficient measures inequality. India's score is 34 (a measure of zero indicates complete equality). In 1970, India's Gini score was 30.2, so inequality is increasing.
- India is ranked at 130 out of 188 countries on the HDI with a score of 0.609.

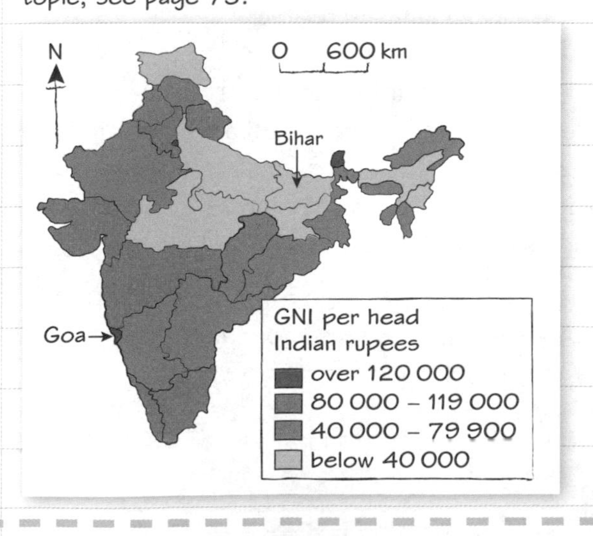

Uneven development

Economic development is also uneven across most LICs and NEEs. In **core** areas, economic development is higher, as is quality of life. In **peripheral** areas, there are fewer opportunities and quality of life is lower. For example, in India:

- Goa is a core state (140 000 rupees GNI per head) with high levels of investment in its developing industrial and service sectors
- Bihar is a state in the periphery (23 435 rupees per head) with a harsh physical environment and few development opportunities.

Has economic development produced inequalities in your case study country?

Now try this

Explain **two** ways in which economic development has had an impact on the environment in your case study country. **(4 marks)**

UK: deindustrialisation, globalisation and policy

You need to know about the causes of economic change in the UK: deindustrialisation and the decline of the UK's traditional industries, globalisation and government policies.

Deindustrialisation

In the 1950s, the UK was a world leader in manufacturing. Around 40 per cent of people in the UK worked in manufacturing. At least 25 per cent of all global manufacturing exports were British. Now only 8 per cent work in manufacturing and 2 per cent of global manufacturing exports are British.

This **deindustrialisation** has happened in many HICs. Traditional industries – steelmaking, shipbuilding, car manufacturing, textile production – have all declined.

Globalisation

Globalisation is the process by which flows of trade and investment build connections between countries. For the UK, globalisation has:

- brought **deindustrialisation** – other countries with lower wages can make things more cheaply
- brought huge investment into London's **banking and finance services** – the UK is a global financial centre.

Government policy

Different UK governments' policies have tried to deal with problems caused by deindustrialisation.

- Nationalisation (after the Second World War) meant state ownership of big industries like steelmaking.
- From the 1980s, the government sold state companies and encouraged new industries. Areas that were not attractive to new industries declined.

Recent governments used new investment and infrastructure to try to boost UK manufacturing.

Worked example

Fact file: London Docklands

- In the 1930s, the London docks were the world's largest.
- Global trade moved to container ships: these ships were too big for London docks.
- Between 1951 and 1981, 100 000 jobs dropped to just 27 000 in the Docklands.
- In 1981, the government invested £1.8 billion on regenerating the Docklands.
- 100 000 new jobs were created in financial services and business services.

Study the Fact file about the London Docklands. Suggest **one** reason why the Docklands were able to recover from deindustrialisation. **(2 marks)**

Government policy was important in the recovery of the London Docklands because the £1.8 billion spend on regenerating the area made it attractive to new financial services and business service companies looking to locate in London.

This is a good answer because it identifies an appropriate reason and explains how it helped the Docklands to recover.

Now try this

Study the photo. It shows workers at Kellingley Colliery, the UK's last deep coal mine, which was closed in 2015. Although the UK still uses coal to produce energy, it is cheaper to import it.

Explain how globalisation is involved in the closure of Kellingley Colliery, Britain's last deep coal mine. **(2 marks)**

UK: post-industrial economy

The UK is now moving towards a post-industrial economy, dominated by service industries and the quaternary sector. New industries have developed, too, which can have much lower impacts on the environment than traditional industries.

As the UK's traditional secondary industries (such as steelmaking and shipbuilding) began to decline, they were replaced by service industries (tertiary industries), including education, transport, health and finance.

Service industries now dominate the UK economy. Over 80 per cent of jobs are in services, and 90 per cent of women work in service industries. Quaternary industries – services based on information technology – have also developed.

Worked example

This photo shows Cambridge Science Park, a cluster of technology and research industries, linked to Cambridge University. Explain why businesses would choose to locate in a science park or business park. **(4 marks)**

There are big advantages for information-based industries to locate in a research environment like Cambridge. There are lots of potential employees with the research and information skills they need. Businesses that work in similar areas benefit from being clustered as they can share services and may be able to form productive partnerships.

Impacts

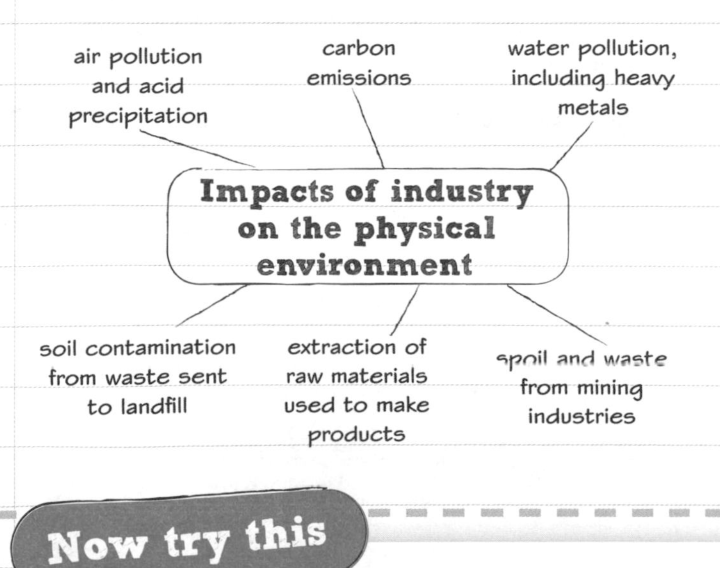

Sustainable industry?

Modern industrial development can be more sustainable, including:

- more efficient processes that use fewer raw materials
- energy-efficient buildings
- reducing waste and recycling materials.

The Range Rover Evoque is made in Liverpool. It is made with 50 per cent recycled materials as part of a strategy to reduce the amount of waste sent to landfill. All metal waste is sorted so valuable aluminium can be recycled.

Now try this

Using an example you have studied, explain how modern industrial development can be more environmentally sustainable. **(6 marks)**

Had a look ☐ Nearly there ☐ Nailed it! ☐

UK: rural change

You need to know about social and economic changes in the rural landscape in **one area** of population growth and **one area** of population decline.

Growing rural areas

The rural environment has many attractions.

- Coastal rural areas are attractive for retired people, especially the UK's south coast.
- Rural areas that are close to urban areas grow due to **counter-urbanisation**: people commuting from rural areas to towns.

Commuter villages have social and economic changes: leisure services increase (golf, horse riding) but village shops may decline because commuters shop in town.

Changes in a declining rural area

Remote upland communities

Where: For example, Snowdonia and Scottish Highlands.

Main employment: farming, mining (primary sector).

Population: decline because of loss of primary industries and **pull factor** of urban areas.

Worked example

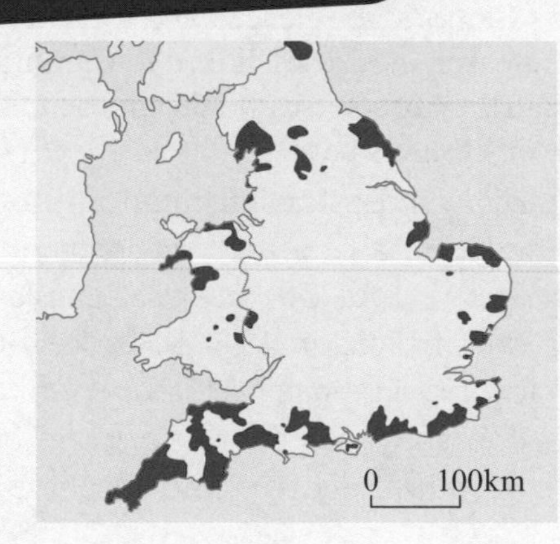

This map shows 'retirement areas' – areas (usually rural) with significantly high proportions of retired people – in England and Wales.
Suggest **one** social and **one** economic change for rural areas that become popular retirement areas.
(2 marks)

Social change: Leisure services for older people will be targeted. These areas may not have much for young people to do or enjoy.

Economic change: There may be an increase in jobs in nursing and caregiving in the area, although wages for these jobs may be low.

Decline in services

Rural areas with declining population often experience a decline in rural services.

- **Post Offices, bank services and shops** – it is expensive to keep these open for small numbers of people. Then people have to travel a long way to access these services, larger in other settlements.
- **Village schools** – it is often too expensive to run small village schools, so children have to be bussed to schools in the nearest town.
- **Bus services** – it can be too expensive to run bus services, which means elderly people can become isolated if they do not drive.

Spiral of decline in rural areas

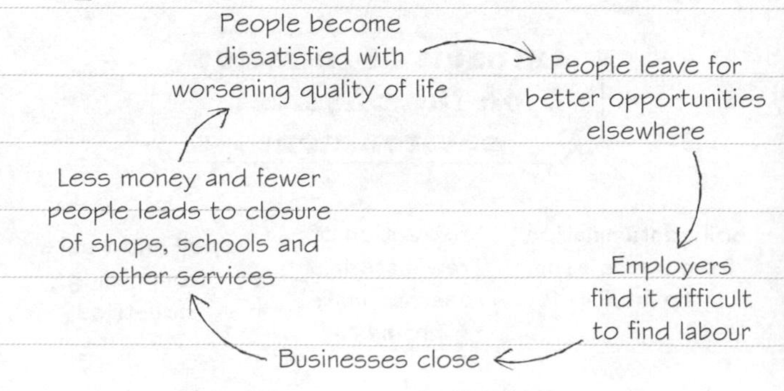

Now try this

Green belts prevent new housing being built in rural areas around major UK cities. Suggest **one** reason why city governments do not want rapid settlement growth in these rural areas. **(2 marks)**

UK: developments

Improvements and developments such as better roads, faster rail links or bigger airports can bring about economic change – and can help resolve regional differences, like the North–South divide.

Road and rail infrastructure

High-speed rail can significantly reduce travel times between different regions of the UK.

Pros and cons to high-speed rail include:

- 👍 More investment in the north, reduced congestion on north–south motorways.
- 👎 London could suck in all the real economic benefits; the cost – HS2's budget in 2016 was £50 billion.

Port and airport capacity

Investment in ports and airports can boost regional development by increasing the amount of trade through ports and allowing greater volumes of people to travel from the region by air. For example, Heathrow, the UK's largest airport, will be expanded.

The North–South divide

The economic growth of London and the south-east of the UK is often contrasted with economic problems in the north linked to deindustrialisation. Government strategies to reduce regional differences in the UK have included:

- creating special investment zones in poorer areas, with infrastructure and tax benefits to attract new businesses
- moving government departments to poorer regions to create more employment
- giving more political power to mayors so they can attract investment to their cities
- investing in new transport infrastructure, such as the HS2 rail network, to speed up travel between north and south.

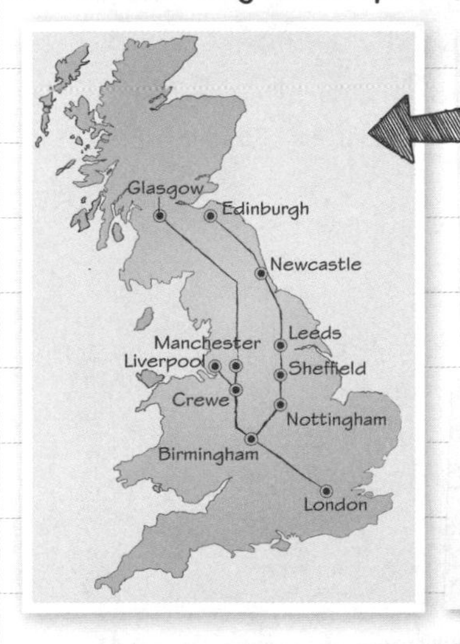

HS2 aims to reduce travel times between London, Birmingham, the East Midlands, Leeds, Sheffield and Manchester. For example, the London–Manchester trip would take an hour less.

House prices are higher in the south because this is where most people want to live and wages are highest.

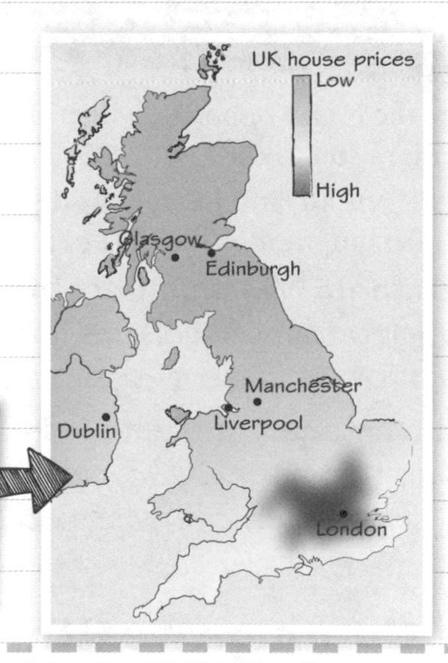

Worked example

Suggest **one** way in which improvements in road infrastructure could reduce regional differences. **(2 marks)**

Making motorway travel faster – for example, by adding a fourth lane – means people in poorer regions have better access to opportunities in richer regions.

Now try this

Study the two maps on this page. Using the maps and your own knowledge, explain how investing in HS2 could reduce the regional differences in UK house prices. **(6 marks)**

Had a look ☐ Nearly there ☐ Nailed it! ☐

UK and the wider world

You need to know about the place of the UK in the wider world – its links to the rest of the world, including its economic and political links to the European Union and Commonwealth.

UK in the wider world

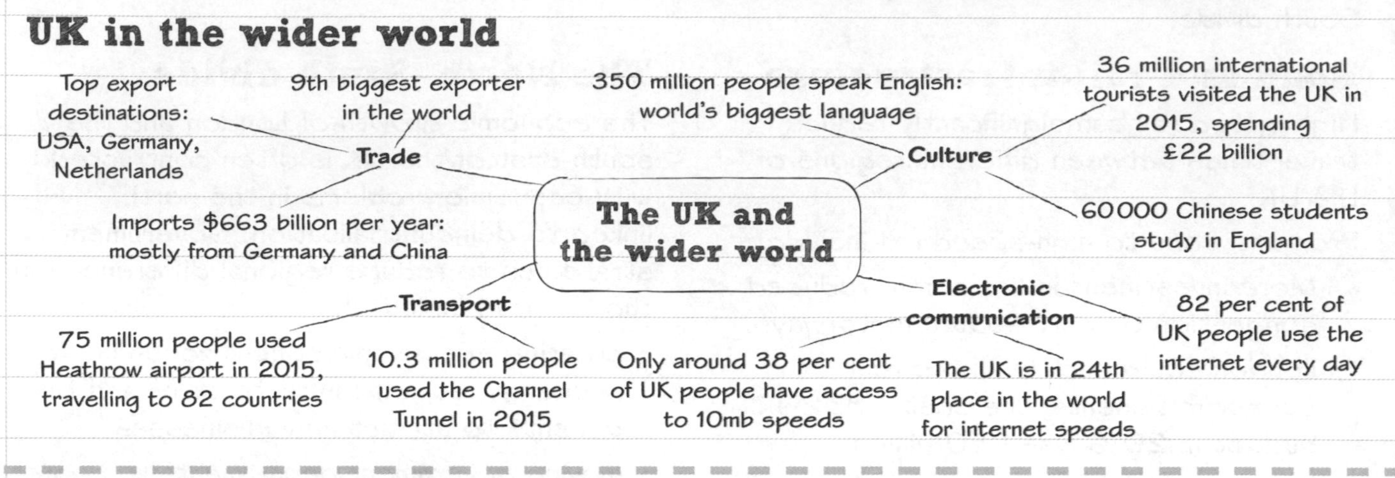

The European Union

The European Union (EU) is an economic and political union that covers most of the European continent. It is the world's largest trading area.

400
300
200
100
0
2000 2005 2010 2015
Total non-EU immigration
Total EU immigration

Immigration into the UK, split into migrants coming from the EU and migrants from the rest of the world

Worked example

Study the graph opposite, which shows EU and non-EU migration to the UK, 2000–2015.
Suggest **one** advantage and **one** disadvantage to the UK of immigration from EU countries. **(2 marks)**

Advantage: Migrants from the EU have been very important as workers in the NHS and in education because of their valuable skills.

Disadvantage: The rapid increase in EU immigration after 2003 meant some UK communities felt overwhelmed by changes.

The Commonwealth

The UK was once the head of an enormous global empire. When this empire broke up through the 20th century, most of its colonies joined the Commonwealth of Nations.

- The Commonwealth is a voluntary intergovernmental organisation. Queen Elizabeth II is the Head of Commonwealth.
- The Commonwealth aims to help its members cooperate in different ways to help each other develop and trade.
- Commonwealth countries share some common links: English language and laws.

There are 52 nations in the Commonwealth. The biggest by population is India (1.26 billion) and the smallest is Tuvalu (10 000). One in three people in the world live in the Commonwealth, but most would not know this!

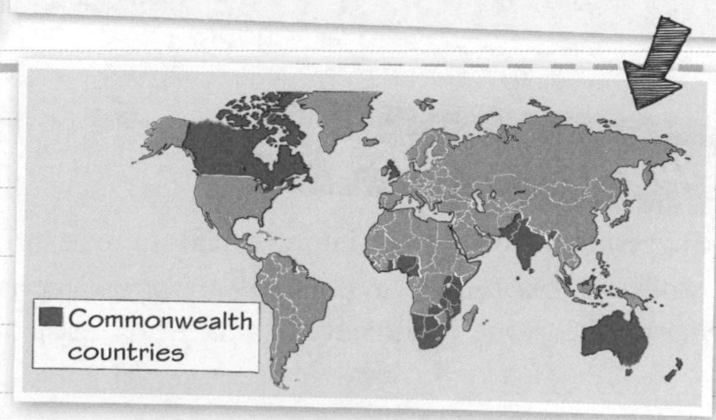

Now try this

'The political disadvantages of EU membership were greater for the UK than the economic advantages.' Do you agree with this statement? Justify your decision. **(9 marks)**

Essential resources

Human development depends on food, water and energy. However, these essential resources are distributed unevenly around the world, which contributes to global inequalities.

Food resources

Humans need to consume around 2000 calories a day to stay healthy. Around the world, 1 billion people do not have enough to eat and 1.9 billion are overweight – around one-third of those are obese. As countries develop, food consumption (calories per person) increases.

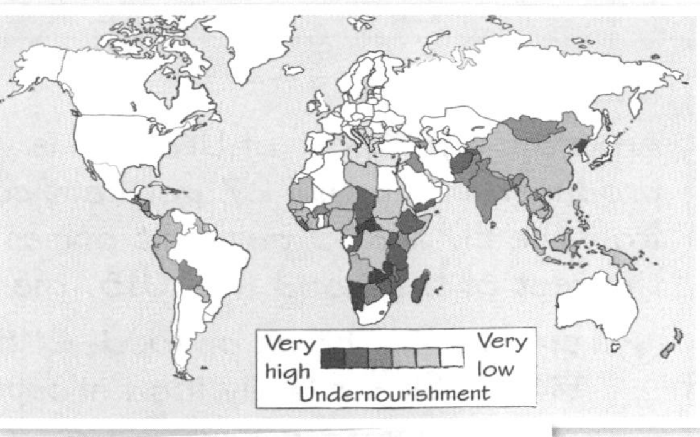

Undernourishment in 2015 – undernourishment has serious health impacts and makes working and studying very difficult

Worked example

Suggest **one** reason why undernourishment is unevenly distributed globally. **(2 marks)**

The most undernourished countries are also the lowest-income countries. People are too poor to reliably get enough food.

Water

Water is essential to life and to economic development. Agriculture needs water for crops and livestock. Industry and energy production rely on water: for example, for cooling.

Many LICs are in areas of water deficit – where low rainfall and high temperatures lead to high rates of evaporation. This impacts on how much food can be grown, and on health, as people have to depend on unsanitary water sources.

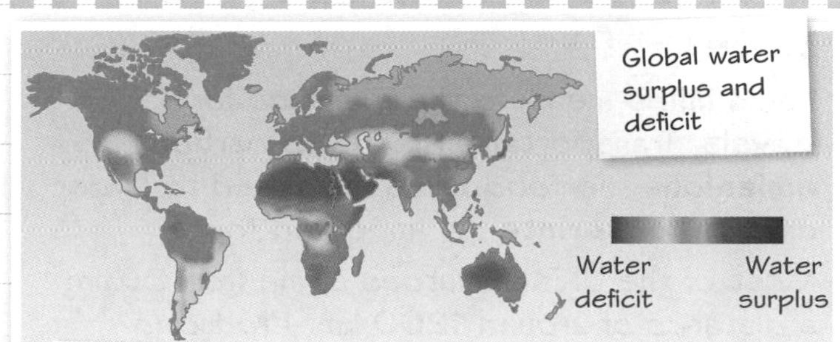

The United Nations estimates that people in sub-Saharan Africa spend 40 billion hours a year collecting water – the same as a whole year of work by the whole of France.

Energy

Global energy consumption is rising: as countries' development occurs, demand for energy increases. The main source of global energy is oil. A few countries have large-scale oil resources, especially in the Middle East, but most do not. HICs dominate global energy consumption but not all HICs have their own energy resources. For example, Japan imports around 84 per cent of the energy it uses.

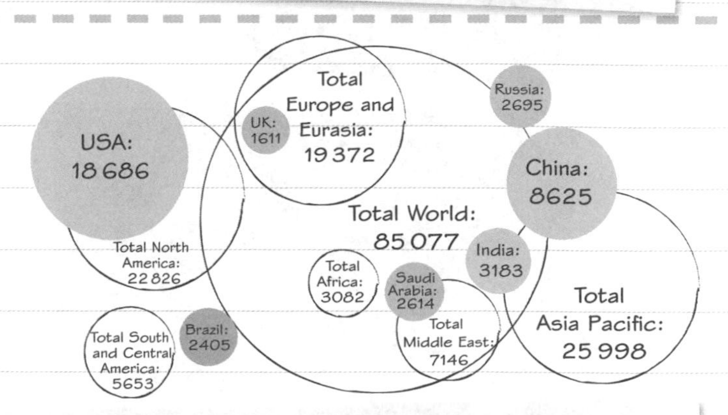

World map of oil consumption by country and region. Each circle represents the number of barrels of oil used per day. The biggest is the USA: over 18 million barrels per day.

Now try this

Explain **one** reason why oil consumption in Africa is lower (3.1 million barrels per day) than in the USA (18.6 million barrels per day), even though 324 million people live in the USA while 1.23 billion people live in the countries of Africa. **(2 marks)**

Had a look ☐ Nearly there ☐ Nailed it! ☐

UK food resources

People in the UK expect to be able to have the food they like all year round, and at affordable prices. The UK imports around 46 per cent of the food products we eat each year.

UK food facts

Around 54 per cent of UK food is produced in the UK, 27 per cent comes from the EU and 19 per cent comes from the rest of the world. In 2015, the UK:

- ✓ spent £198 billion on food, £115 billion of it for family food shopping
- ✓ imported £38.5 billion of food and exported £18 billion
- ✓ spent £8.5 billion on 'ethical' foods (organic and fair trade foods).

All-year round supply

The UK imports 77 per cent of its fruit and vegetables from more than 24 countries.

- Not all fruit and vegetables can be grown commercially in the UK – like bananas.
- Most fruit and vegetables are **seasonal** – they only grow at certain times – so must be sourced from different places year round.

Bananas do not grow in the UK, but are the UK's favourite fruit

Carbon footprints and food miles

Food miles measure the distance food travels. Transportation leads to **carbon emissions**, so food miles are linked to larger **carbon footprints** for imported food.

Most of the UK's tomatoes come from Spain: a distance of around 1200 km. Producing tomatoes out of season in the UK means much higher carbon emissions than importing them from Spain because UK growers would have to use artificial heat and light.

Agribusiness

Just 1.5 per cent of the UK's workforce works in farming. The UK can produce 54 per cent of all the food it consumes with only a small workforce because most production is by agribusinesses – large corporations running highly mechanised and efficient farming operations. Many are located in East Anglia.

Between 2007 and 2015, food prices in the UK increased by 34 per cent, mainly due to increases in global agricultural prices.

Worked example

The photo shows a farm shop that sells food and drink produced in the local area. Suggest **one** reason why sourcing food locally could reduce the carbon emissions associated with food production.

(2 marks)

If people buy locally produced food then this will reduce the distance food is transported, reducing the food miles. However, this will only reduce carbon emissions for fruit and vegetables if they are only sold when they are in season in the UK.

Now try this

Outline **one** reason why people in the UK might choose to buy 'ethical' foods. **(2 marks)**

UK water resources

As the population of the UK increases, the demand for water also increases. The areas with the highest rain are not the areas with the highest population, so supply must be managed.

Water usage in the UK

- Farming and industry use a lot of water, but most is used as coolant in the power industry.
- Rivers and pipelines are used to transfer water from major reservoirs in wet areas to big cities. However, pipelines are expensive.
- Areas in the south and east have smaller reservoirs, and rely on groundwater supplies. Drier areas and highly populated areas can suffer water stress.

Areas of **surplus**: areas with lots of rain and low populations
Areas of **deficit**: drier, highly populated areas

Causes of pollution

1. Farmland runoff: fertilisers and pesticides
2. During storm events, untreated sewage sometimes has to be flushed into rivers
3. Runoff from roads and motorways, including salt used for gritting in winter

Managing UK water quality

- Strict laws set out limits for water pollution and punishments for those that break them.
- Lower water levels increase concentration of pollutants.
- Laws control how much water is taken (abstracted) from rivers and groundwater.
- Efficient treatment plants process sewage and recycle safe water back into rivers.

Worked example

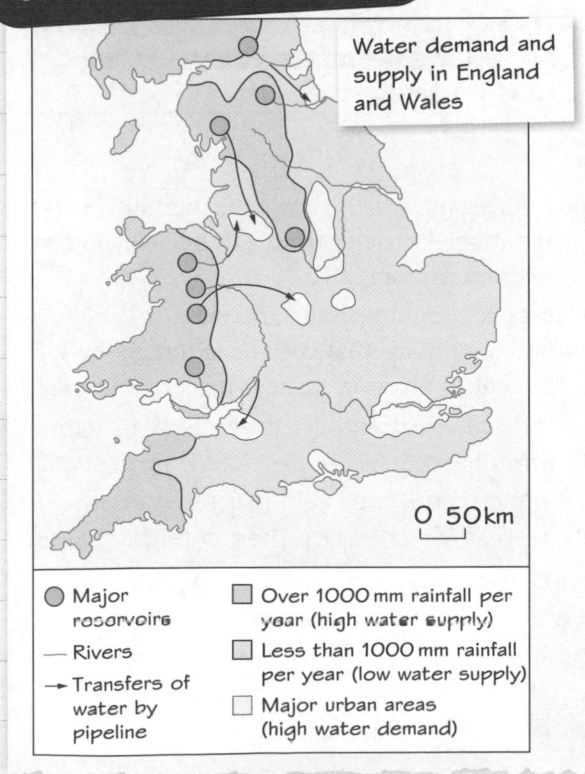

Study the map. **1** Suggest **one** reason why the demand for water in the UK is rising. **(1 mark)**

As the populations of major cities like London and Birmingham increase, there is more demand for water for household use.

2 One solution in the south and east of the UK would be large-scale water transfers from areas of water surplus to areas of deficit. However, water transfer schemes have problems.

Explain **two** problems with large-scale water transfer schemes. **(4 marks)**

Water is very heavy to transport. While it will flow downhill, it needs to be pumped uphill. Pumping uses a lot of energy, which is likely to mean an increase in carbon emissions.

The easiest way to transfer water is to link rivers. For example, the Severn River (water surplus) could be linked to the River Thames (deficit). However, the two rivers have different water chemistry. Water from the Severn could damage wildlife in the Thames.

Now try this

Explain why an increasing demand for energy is linked to an increasing demand for water. **(2 marks)**

Had a look ☐ Nearly there ☐ Nailed it! ☐

UK energy resources

The UK's energy mix is changing: although the country still relies heavily on fossil fuels to generate the energy it needs for homes, transport, industry and farming, this reliance is decreasing.

The changing energy mix

Historically, the UK relied on its extensive coal resources for energy. In the 20th century, the UK developed oil and gas production in the North Sea. However, different factors have meant changes to the UK's energy mix.

- It became cheaper to use coal from other countries and many UK coal mines closed.
- Oil and gas reserves in the North Sea have begun to run out.
- Other countries supplying the UK with energy are not always politically stable.
- Concern over climate change means a reduction in the use of fossil fuels and an increasing role for renewable energy as part of the UK's energy mix.

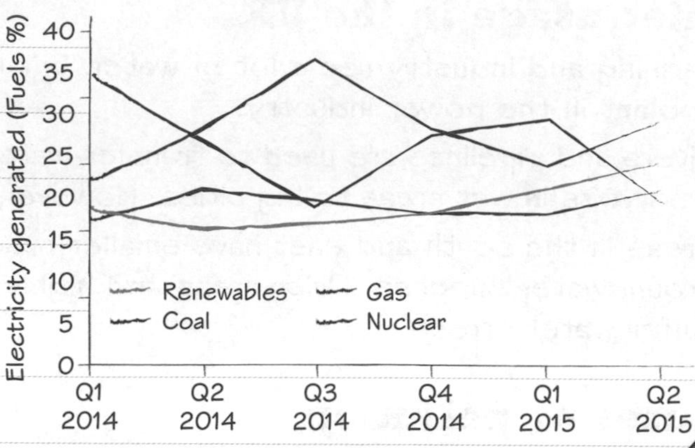

This line graph shows that, in the first three months of 2015, more renewable energy was used to generate electricity than coal for the first time in the UK. Renewable energy is less efficient at generating power than fossil fuel sources but carbon emissions are far lower.

Fracking – shale gas extraction

There are concerns about fracking's possible environmental risks

There are likely to be economic benefits from an extension of fracking in the UK. Jobs will be created in the industry, and the UK will have increased **energy security**, and will be able to reduce its reliance on foreign energy suppliers.

However, fracking has environmental risks. It involves pumping water and chemicals into the shale bedrock, which could contaminate groundwater. Shale gas is also a fossil fuel so using it for energy production would not help the UK to reduce carbon emissions.

Worked example

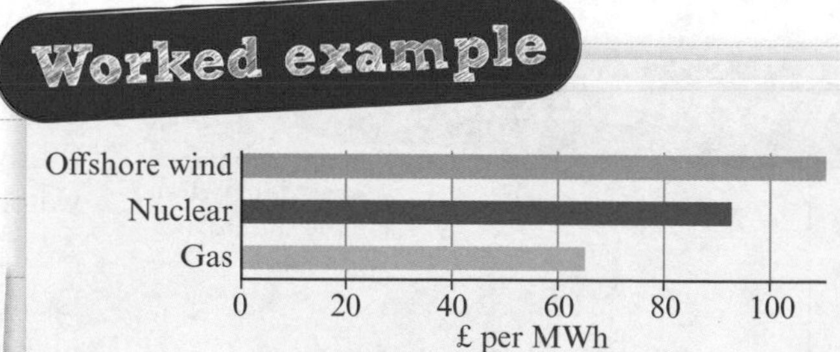

Study the graph, which shows the different costs of generating electricity from gas, nuclear power and offshore wind in 2016.

Suggest **one** economic issue and **one** environmental issue that could encourage the UK government to invest in nuclear power. **(2 marks)**

Economic issue: Although nuclear power is approximately £28 per MWh more expensive than gas, it is approximately £16 per MWh cheaper than offshore wind.

Environmental issue: Nuclear power has very low carbon emissions, while gas is a fossil fuel. So nuclear power would help the UK meet its international targets for lowering CO_2 emissions.

Now try this

Suggest why fracking could increase the UK's **energy security**. **(2 marks)**

Demand for food

Food security is about people always having access to enough safe and nutritious food for them to have a healthy and active life. More than 795 million people in the world are **food insecure. Food, Water and Energy are options: only revise the one you studied.**

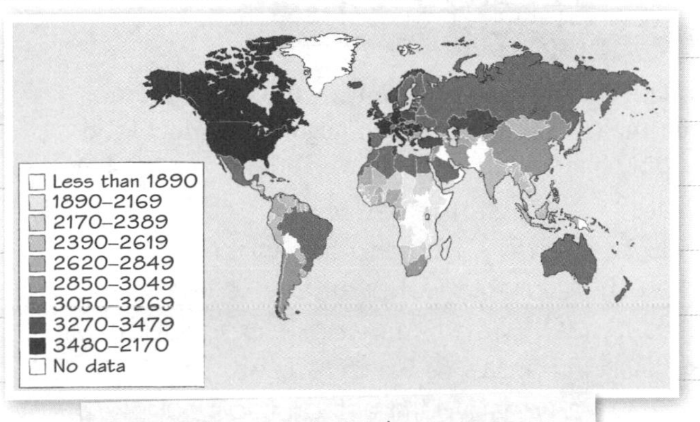

Map of daily calorie intake per person. Food supplies us with energy, measured in kilocalories: **kcals.**

Global patterns of calorie intake

- HICs have the highest calorie intake: 3420 kcal per person per day on average.
- Many NEEs have seen a rapid increase in calorie intake per person. In the last 10 years, the average intake for China has increased from 2500 to 3000 kcals per person.
- LICs often have lower calorie intakes. The average is 2240 kcal per person per day.
- The poorest countries in central Africa have an average of 1820 kcal per person per day.

Global food production

This table shows the amount spent by different world regions on exporting and importing food in 2012. The figures are in US $millions.

	Export value	Import value
Europe	403	418
Asia	160	264
Americas	266	159
Africa	29	64
Oceania	45	14

Global food supply

HICs can supply a lot of food to their populations. Some HICs, like the USA, are major food producers. All HICs can afford to import food they cannot produce themselves, so consumers have a wide range of affordable food.

Although high percentages of LIC populations farm for a living, food production is often low. People produce what their families need to survive, and little more. Food exports and imports are often low. Most people can only afford a very basic diet.

Worked example

Study the graph, which shows changes in daily calorie intake in different world regions over time.

Suggest **one** reason why economic development increases calorie intake. **(2 marks)**

As people earn more, they can afford to buy more food, and a wider range of foods including more meat and higher calorie foods.

There are two reasons why more food is being eaten: a rising population (more people to feed) and economic development.

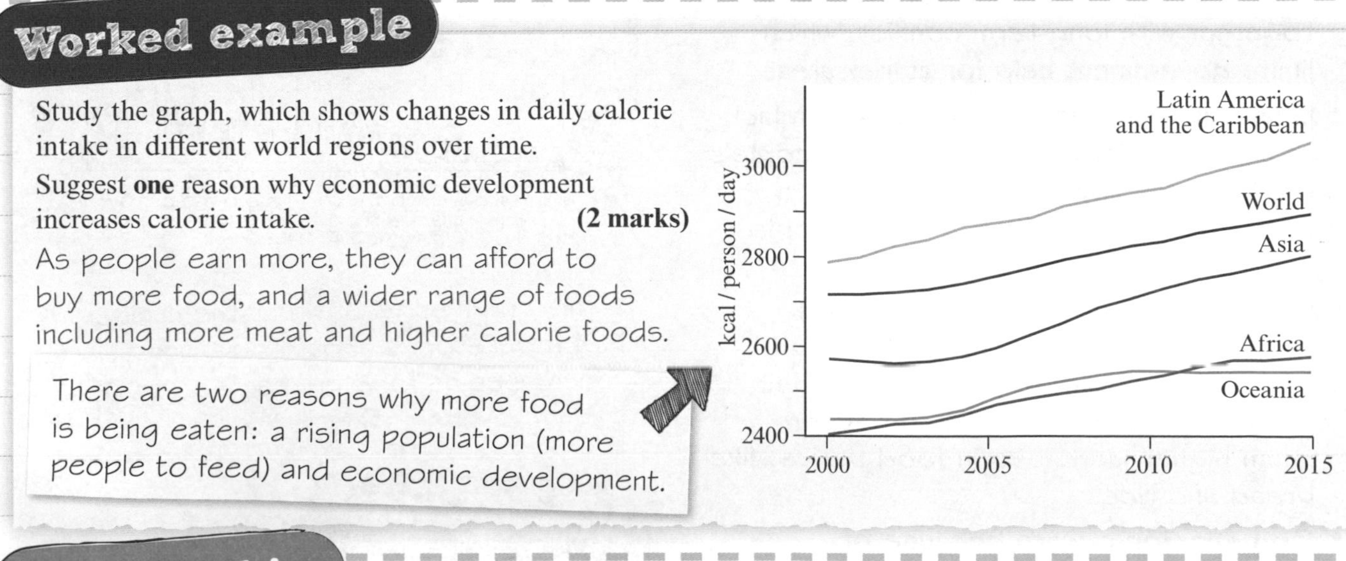

Now try this

Suggest another reason why economic development increases calorie intake per person. **(2 marks)**

Had a look ☐ Nearly there ☐ Nailed it! ☐

Food insecurity

Different factors affect food supply. Lack of reliable access to enough safe, nutritious food causes food insecurity, which has very serious and negative impacts for people and the environment.

Factors affecting food supply

The diagram below lists the main factors affecting food supply. They are often linked.

For example, climate change is making some areas too dry for the crops traditionally grown there. Increased use of irrigation creates water stress. As crops fail, poverty increases. Conflict arises as people compete for water.

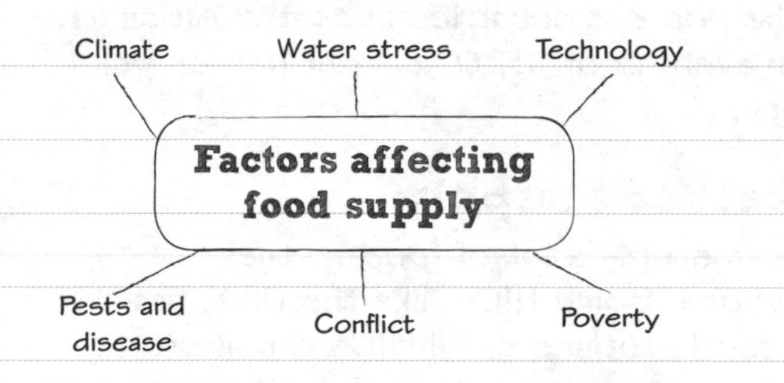

Worked example

Using the diagram and your own understanding, outline **two** factors that can negatively affect food supply in a country. **(4 marks)**

Water stress can negatively affect food supply. This is when the demand for water is greater than the available water supply. It is estimated that by 2030, half the world's population will be living in areas of water stress. This is likely to have a negative impact on food supply, especially for crops that need irrigation.

Poverty is another factor that can affect food supply. If people cannot afford to buy equipment, fertilisers and disease-resistant seeds, their crop yields are often low. This means not enough food to eat for their families and not enough money to buy more food.

Impacts of food insecurity

- **Famines** occur when at least 20 per cent of households in an area face extreme food shortages, when more than 30 per cent of people are acutely malnourished and when more than two people per day per 10 000 people are dying. Famines are rare and are often caused by long-term drought together with long-term conflict, which limits government help for at-risk areas.
- **Undernutrition** is when a long-term lack of enough nutritious food means people's physical development and health is affected. Undernutrition means people can become weak and unwell.
- **Price increases** happen when food becomes harder to access. Although poor farmers can earn more from their crops, this is not enough to pay for much higher increases in food basics, like bread and rice.
- **Soil erosion** occurs because of overgrazing by livestock, deforestation to produce more farming land, overuse of soils and farming marginal areas.
- **Social unrest** in the world today is often linked to droughts, reduced food supplies and higher food prices.

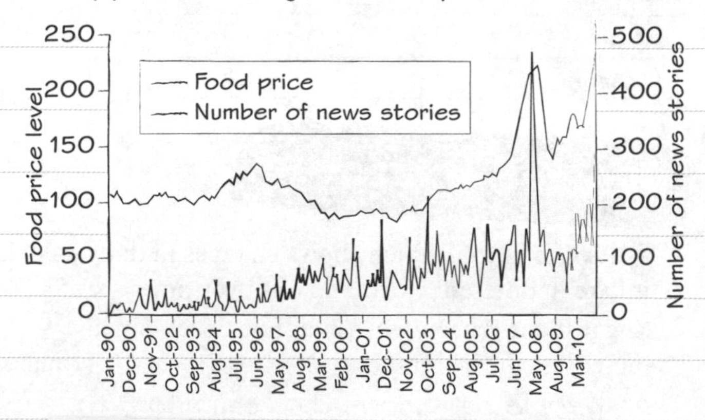

This graph suggests that the number of news stories about social unrest increases as food prices rise

Now try this

Suggest how conflict can affect food supply. **(4 marks)**

Increasing food supply

As well as having an overview of different ways to increase food supply, you need to know an **example** of a large-scale agricultural development to show its advantages and disadvantages. One has been suggested here, but **revise the example you did in class if it is different.**

Key terms

Irrigation: artificially adding water to the land to help grow crops.

Aeroponics: growing plants by suspending them in the air, with nutrients delivered to their roots in a fine mist of water; excess water is collected and reused.

Hydroponics: growing plants just in sand, gravel or liquid, with nutrients added.

Biotechnology: includes the genetic modification of plants and animals to give them additional characteristics: for example, resistance to a common disease.

Appropriate technology: technology at the right level to be useful and accessible to the communities who are using it.

Horticulture in Kenya

Horticulture is big business in Kenya, an LIC in east Africa. Kenya's horticulture exports are worth US$1 billion a year. The farms are modern large-scale agribusinesses and use sophisticated technology to maximise profits, including drip-feed irrigation, artificial lighting and 'fertigation': adding fertilisers to irrigation.

- Advantages: Horticulture contributes 1.4 per cent to Kenya's GNI; 100 000 Kenyans work directly in the industry.
- Disadvantages: Most employees are low-paid; farms have taken up most of the fertile land; environmental concerns about fertiliser pollution; vegetables, fruit and flowers have a high carbon footprint.

Advantages

- Irrigation means crops can be grown in regions that would otherwise be too arid.
- Aeroponics and hydroponics use less water, reduce disease, increase yields.
- Biotechnology gives crops valuable qualities: added nutrition; resistance to drought, pests and disease; longer shelf-life.
- Appropriate technology means people get solutions that directly meet their needs.

Disadvantages

- Irrigation can use up groundwater, lower river levels and encourage salinisation.
- Aeroponics and hydroponics are expensive. In the UK, they use energy for heat and light to grow crops out of season.
- Biotechnology can be expensive for farmers if companies engineer seeds so they need to be bought again each year.
- New technologies can demand high skill.

The new green revolution

The new green revolution aims to use biotechnology to develop seeds for specific conditions, such as resistance to drought. The technology will also make crops more nutritious and increase yields. There are concerns, though, that the seeds will be engineered so that farmers must buy new seeds each year.

Worked example

Explain **one** way in which appropriate technology can increase food supply. **(2 marks)**

An example is the Universal Nut Sheller: a hand-operated mill that can shell 50 kg of peanuts per hour. This technology costs $50 to make, lasts 25 years and is easy to maintain. A community can process their peanut crop much faster, so they are encouraged to grow a larger crop.

Now try this

Describe the advantages and disadvantages of the example of a large-scale agricultural development that you have studied. **(6 marks)**

Had a look ☐ Nearly there ☐ Nailed it! ☐

Sustainable food supplies

Most of the ways we currently produce food are not sustainable. Sustainable food supplies do not overuse natural resources, so future generations will be able to use those same resources to produce their food, too. You need an **example** of a local scheme in an LIC or NEE to increase sustainable supplies of food.

Key terms

Organic farming: uses no artificial chemicals and aims to protect the environment and biodiversity through sustainable resource use.

Permaculture: is farming that follows the way natural ecosystems work: a 'forest garden' approach that does not deplete resources.

Urban farming: involves growing, processing and delivering food where most people live: in cities and towns.

Sustainable meat: is meat from animals that have been pasture-fed, as part of a natural ecosystem. **Sustainable fishing** means making sure that future generations will be able to fish the same species in the same way.

Seasonal food consumption: means only eating fruit and vegetables that are in season.

A rooftop farm in New York City, USA

Advantages and disadvantages

- Organic farming maintains soil fertility but organic crops cost more to produce.
- Permaculture: pollution-free and very little waste (all used in compost), but its slow methods do not suit large-scale production.
- Urban farming: reduces food miles and encourages more nutritious diets, but crops are affected by urban pollution.
- Sustainable meat is healthier meat with less fat and fewer calories. However, it is more expensive to rear and that makes it more expensive for consumers to buy.

Example

Rice-fish farming

Rice-fish farming is a way to increase sustainable supplies of food. It is used in many Asian LICs and NEEs, including China, India and the Philippines.

Because rice seeds need 10–20 cm of water to grow in, it is possible to farm fish along with rice. This increases sustainable supplies of food because:

- farmers can eat the fish, adding protein to their diet, or sell the fish for added income
- the fish eat pests and weeds, and fish excrement fertilises the rice. Farmers do not need to spend money on pesticides and fertilisers.

Reducing waste

The UK wastes up to 10 million tonnes of food and drink a year. Considering all the resources that go into producing this food and drink, and the carbon emissions involved, reducing waste is a good way to start moving towards more sustainable supplies of food.

Worked example

Suggest **one** way seasonal food consumption can increase sustainable supplies of food. **(2 marks)**

Currently, when fruit and vegetables are out of season in the UK, they are shipped or flown in from other countries. Only eating the fruit and vegetables that are available locally in each season cuts back on food miles.

Now try this

Explain how a local scheme in an LIC or NEE that you have studied has increased sustainable supplies of food. **(6 marks)**

Demand for water

Water security is about people having reliable access to enough safe water for them to have a healthy and productive life, and about reducing the risk of water hazards.

Water, Food and Energy are options: only revise the one you studied.

Global patterns of water surplus and deficit

This map is of projected areas of **water surplus** (little or no water scarcity) and **water deficit** (water scarcity) in 2025.

- Climate change is likely to lead to longer periods of drought and less predictable rainfall across the global 'South'.
- Economic water shortage is when there is not enough investment in infrastructure to supply enough water to people, rather than a physical shortage of water.

Compare this map with the map of current water surplus and water deficits on page 103.

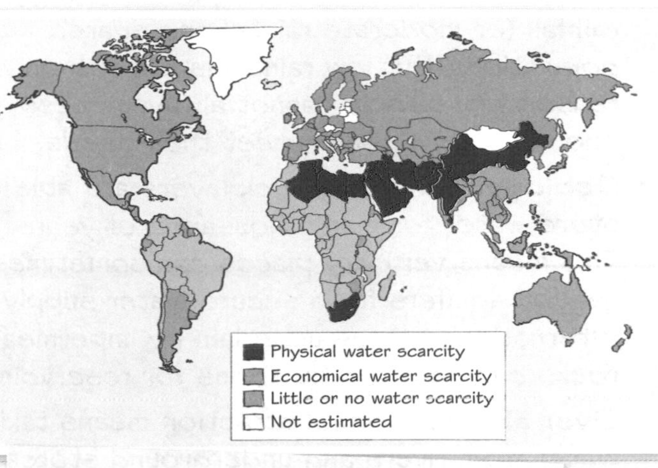

Projected water scarcity by 2025; climate and population growth will continue to be very significant factors in increasing water scarcity

Why increase water consumption?

1. Rising population
2. Economic development

Rising population numbers puts more pressure on fresh water supply. The rate that water consumption is increasing is **twice** the rate of population growth. More people does not just mean more drinking water. Water is also essential for growing food for rising populations, and for **economic development**.

Key terms

Water-stressed regions: have less than 1700 m³ of water per person per year.
Water-scarce regions: have less than 1000 m³ per person per year.

Amount of water needed to:

- ✓ grow enough cotton to make one pair of jeans = 6800 litres
- ✓ grow 1 kg of rice = 2500 litres
- ✓ raise 1 kg of beef = 15 000 litres
- ✓ make 1 tonne of steel = 237 000 litres
- ✓ make 1 hour of electricity = 95 litres, per household.

Worked example

Using the information opposite and your own knowledge, outline **two** ways in which economic development increases water consumption. **(4 marks)**

One reason is that, as countries develop economically, people get wealthier and water consumption increases: for example, because more families buy washing machines. Secondly, as economies industrialise, water consumption increases because industry and energy production need large amounts of water: for example, 1 ton of steel requires 237 000 litres of water.

Now try this

Suggest **one** reason why a country can go from being in water surplus to water deficit. **(2 marks)**

Had a look ☐ Nearly there ☐ Nailed it! ☐

Water insecurity

You need to know about the factors affecting water availability and the impacts of water insecurity.

Factors affecting water availability

- **Rainfall:** areas of water surplus have high rainfall (or moderate rainfall and sparse population). But low rainfall is not the only reason why people cannot always access enough safe water to meet their needs.
- **Geology:** permeable rock layers are able to store water for many thousands of years. This means very dry places can sometimes rely on aquifers for a secure water supply. Alternatively, areas underlain by impermeable rocks can be good locations for reservoirs.
- **Over-abstraction:** abstraction means taking water from rivers and underground stores. Over-abstraction happens when too much water is taken. For example, if too much water is taken from rivers for irrigation, the river can no longer supply other users.
- **Pollution of supply:** rivers and groundwater can become polluted by industry, agriculture and human waste. People then have to risk illness to use the water.
- **Poverty:** people in LIC and some NEE cities often have to choose between unsafe water and expensive water. Poverty reduces access to water in these cases.
- **Limited infrastructure:** when there has not been the investment in pipes and pumping stations in a region that are needed to get water to those who need it.

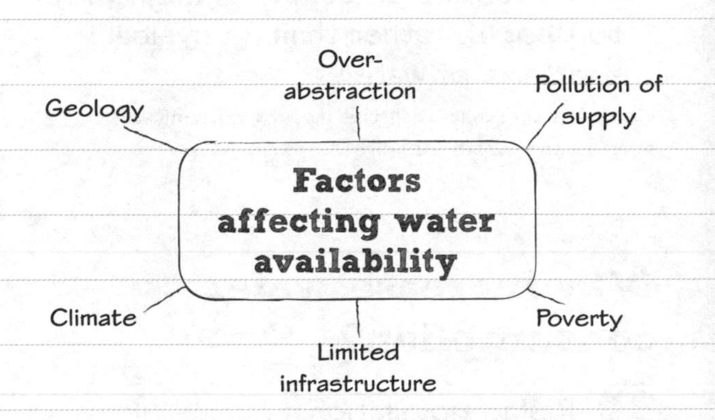

Worked example

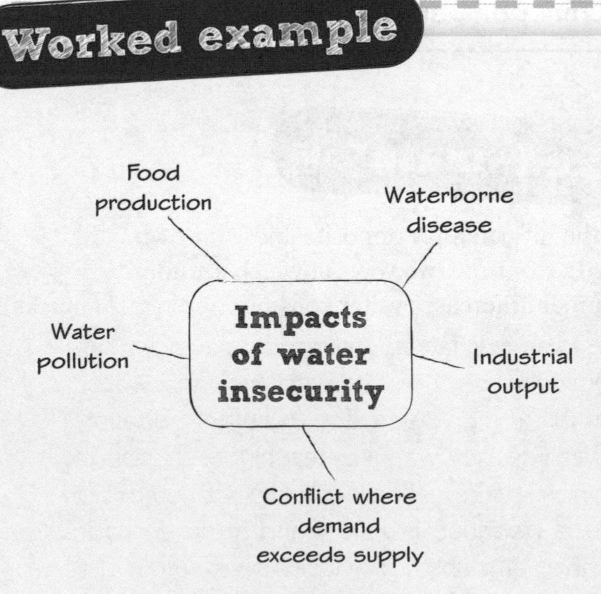

This diagram shows different impacts of water insecurity. Using the diagram and your own knowledge, describe how **two** of these impacts are affected by water insecurity. **(4 marks)**

Impact 1: Waterborne diseases, like cholera, dysentery and typhoid, are spread by water. When people do not have access to clean and safe water supplies – water insecurity – these diseases cause high levels of illness and 3.4 million deaths per year.

Impact 2: Some key industries, such as steelmaking, chemicals and textiles, require water in very large quantities. Where water prices are high due to water insecurity, industrial output will be negatively affected.

Now try this

Pick another **two** impacts from the diagram and outline how they are affected by water insecurity. **(4 marks)**

Increasing water supply

As well as an overview of ways to increase water supply, you need to know an **example** of a large-scale water transfer scheme to show how its development has advantages and disadvantages. One has been provided here, but **revise the one you did in class if it is different.**

Key terms

Aquifer recharging: surface water is diverted into natural **aquifers** (permeable rock layers) to fill them up again.

Dams block rivers to create **reservoirs**. Cities are often supplied in this way (for example, Las Vegas, USA).

Water transfers via pipelines or rivers move water from areas with a water surplus to areas with a water deficit.

Desalination: this process removes salt from seawater.

Desalination membranes filter out salt from seawater

Advantages and disadvantages

- Dams and reservoirs store water for when it is needed, and reduce river flooding. However, in warm climates evaporation means reservoir levels fall quickly unless they are recharged by rainfall.
- Water transfers redistribute water from surplus areas to deficit areas. However, they are very expensive, have environmental impacts and encourage unsustainable water use in areas of naturally lower rainfall.
- The ability to turn saltwater into fresh water offers huge opportunities for human development. But desalination requires a lot of energy and is very expensive to set up.

The South-to-North Water Transfer Project

The SNWTP aims to transfer 12 trillion gallons of water 1000 miles from China's wetter south to the urbanised, industrialised and drier north.

Advantages

- Water security for 500 million people
- Increased opportunities for economic development in the north

Disadvantages

- Enormous cost: US$62 billion
- Major water pollution problems from farmers and factories along the SNWTP route
- Could mean that the north continues to use water unsustainably with overuse of water and water waste

Worked example

Lake Mead is the USA's largest reservoir. Formed by the construction of the Hoover Dam on the Colorado River, it supplies water to 20 million people in three US states, as well as to huge areas of farmland in this semi-arid region of the USA.

Suggest **one** advantage and **one** disadvantage of using a dam and reservoir to increase water supply in an area of water deficit. **(2 marks)**

One advantage is that a large store of water is created, far more than a river can supply at one time, which allows significant economic development of the region.
One disadvantage is that a long drought can lower reservoir levels and put the whole area and its unsustainable economic development at risk of water scarcity.

Now try this

Describe and explain the advantages and disadvantages of the large-scale water transfer scheme you have studied. **(6 marks)**

Had a look ☐ Nearly there ☐ Nailed it! ☐

Sustainable water supplies

Large-scale strategies to increase water supply have generally proved unsustainable because they encourage the overuse and waste of water. Sustainable water use aims to ensure that future generations will be able to use water resources in the same way as we use them today. You need an example of a local scheme in an LIC or NEE to increase sustainable supplies of water.

Key terms

Water conservation: ways to reduce water consumption by using water more efficiently and reducing wasteful use. An example would be washroom taps that shut off automatically.

Groundwater management: this aims to stop over-abstraction of groundwater by measures like setting limits for farmers and encouraging them to use more efficient irrigation systems. Recharging aquifers is another option.

Water recycling: water that has been used by homes and businesses can be reused for other purposes, such as irrigation or cooling for industry or electricity generation, or returned after treatment to remove pollutants to stores, such as reservoirs or groundwater.

'Grey' water: this is water that has already been used for one purpose (for example, washing) and is then used for another (for example, irrigation).

Effluent is water that has been used in the sewage system – a type of water recycling

Advantages and disadvantages

- Water conservation can cut costs: people pay less for using less water. However, it is unpopular when it means people cannot have green lawns or clean cars.
- Groundwater management is an efficient way of storing water for when it is needed, as there is no loss through evaporation. However, it is difficult and expensive to enforce laws on groundwater use.
- Water recycling and use of 'grey' water works at small and large scales: for example, watering gardens with recycled bathwater. However, first people have to be convinced that 'grey' water is safe to use.

Increasing sustainable water supplies

Appropriate technology is often used to increase water supplies, such as sand dams. People in LICs in eastern Africa often lack water during the dry season, when rivers are dry. However, some water is stored in the sand of river beds – a traditional source of water for centuries.

Sand dams enhance this natural process. They are low concrete walls built into river beds. They trap much more sand behind them, forming a much larger store for water. Several sand dams in local river beds means a more reliable dry season water supply for local people.

Worked example

Outline **one** way in which UK households could use water more sustainably. **(2 marks)**

Installing low-flow shower heads: these mix water and air together so that less water is used, but people still experience a 'power shower' effect.

Now try this

Explain how a local scheme in an LIC or NEE that you have studied has increased sustainable supplies of water. **(6 marks)**

Demand for energy

Countries with an **energy deficit** consume (use) more energy than they can generate themselves. Countries with an **energy surplus** create more than they use. **Energy security** is about people and industries having reliable access to enough energy at affordable prices.

Energy, Water and Food are options: only revise the one you studied.

Global energy deficits and surpluses

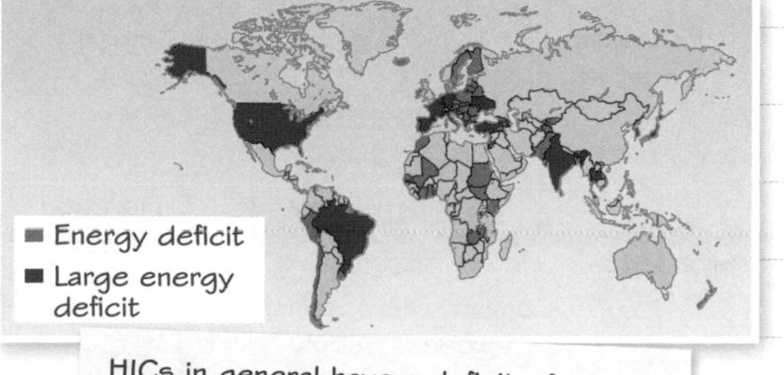

HICs in general have a deficit of energy

Energy surplus
Large energy surplus

Many LICs have a surplus of energy

Reasons for increasing energy consumption

1. Economic development
2. Population growth
3. Technology

Economic development: as countries develop, energy consumption increases. For example, China's energy consumption has increased because of its rapid industrialisation and urbanisation. Industry demands much more energy than agriculture. Also, as people become wealthier, they save to buy a car.

Population growth: more people in a country's population means a rise in demand for energy. For example, between 1950 and 2000, China's population grew from 600 million to 1.3 billion.

Technology: increasing use of technology, such as mobile phones, computers, TV and lighting, all increases energy consumption. The biggest energy users are fridges, freezers, humidifiers, air conditioners, water heaters and central heating systems, which often run permanently.

Worked example

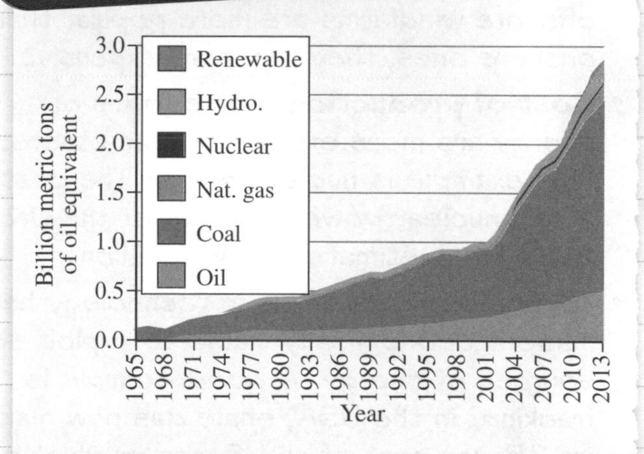

Study the graph above which shows China's energy consumption between 1965 and 2013. Which source of energy makes up the biggest share of China's energy consumption?

- ☐ **A** Hydro-electricity
- ☐ **B** Natural gas
- ☒ **C** Coal
- ☐ **D** Oil

Now try this

1 Using the compound line graph on this page, calculate the contribution of natural gas to China's energy consumption in 2013. **(1 mark)**

2 In 1980, China spent $600 million importing oil. In 2016, it spent $198 billion importing oil. Suggest **two** reasons that help explain this major increase in oil imports. **(4 marks)**

Energy insecurity

You need to know about the factors affecting energy supply and the impacts of energy insecurity.

Factors affecting energy supply

Factors affecting energy supply: Technology; Physical factors; Political factors; Cost of exploitation; Cost of production

- **Physical factors:** a country's geology determines whether it has fossil fuels, and how easy those fossil fuels are to access. Some renewables depend on physical factors too, for example, climate for solar.
- **Cost of exploitation:** geology makes some sources of fossil fuels very expensive to extract: for example, oil fields that are in very deep parts of the oceans. While offshore windfarms are more popular than onshore ones, they are more expensive.
- **Cost of production:** some forms of energy are more expensive than others. One example is nuclear power. The cost of a new nuclear power station for the UK in 2015 was estimated at £18 billion.
- **Technology:** advances in technology have made it economically viable to exploit new sources of energy. A good example is fracking: in the USA, shale gas now makes up 35 per cent of all US gas production.
- **Political factors:** there are many different kinds of political factors affecting energy security. For example, many countries depend on Russia for energy supplies. Russia can threaten to cut off supplies.

Worked example

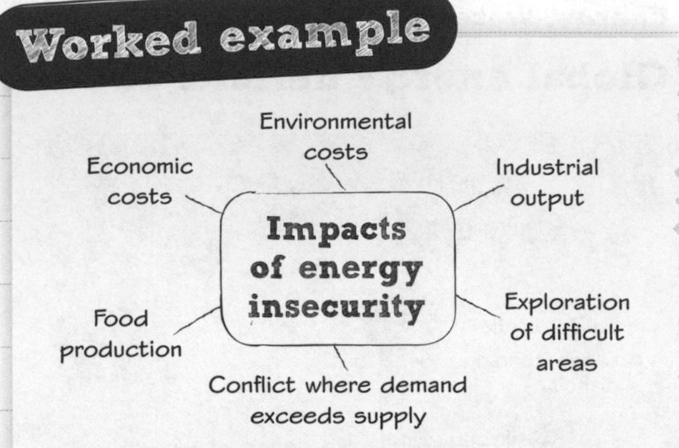

This diagram shows different impacts of energy insecurity. Using the diagram and your own knowledge, outline how **two** of these impacts are affected by energy insecurity. **(4 marks)**

Impact 1: Food production uses a lot of energy – for machinery, processing and transportation – so food production is vulnerable to energy price rises.

Impact 2: All industries need energy, but some industries are very energy-intensive. For example, steel production requires large amounts of energy and often locates where energy is cheapest.

Changes in oil prices are usually triggered by conflicts, or economic or political crises.

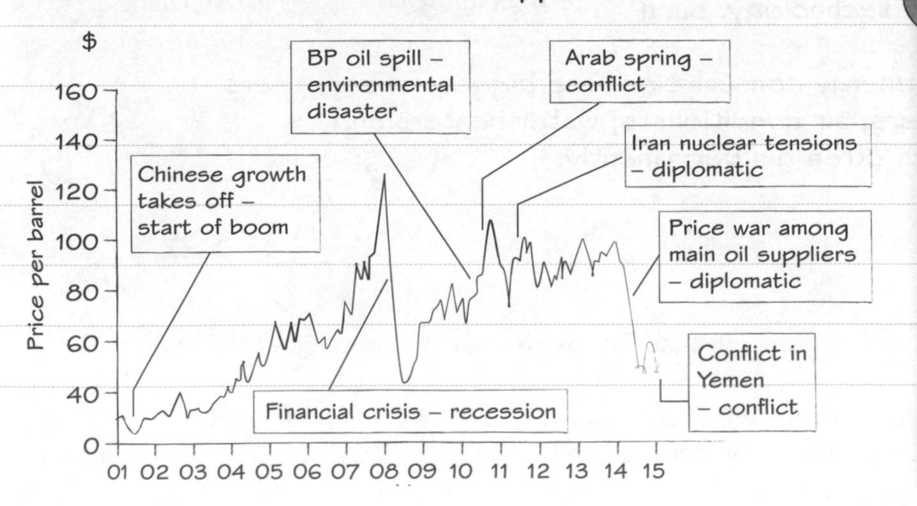

Now try this

1. Pick another **two** impacts from the Worked example diagram and explain how they are affected by energy insecurity. **(4 marks)**
2. Explain how energy insecurity could encourage companies and governments to invest in extracting oil and gas from challenging environments, such as the Arctic. **(4 marks)**

Increasing energy supply

As well as an overview of ways to increase energy supply, you need to know an **example** to show how the extraction of a fossil fuel has advantages and disadvantages.

Key terms

Renewable

Biomass: organic matter used as fuel, includes burning organic matter and processing plants into biofuels.

Wind: wind turns turbine blades to generate electricity.

Hydro: flowing water turns turbines to generate electricity; China leads the world, generating 850 billion kilowatt hours in 2012.

Tidal: a type of hydro power that uses the power of tides to generate electricity.

Geothermal: heat from the Earth is used to heat houses; steam from geothermal heat is used to enable electricity.

Wave: wave energy converters turn the movement of water into power.

Solar: the energy from the Sun can be converted into power through different technologies, including photovoltaic cells.

Non-renewable

Fossil fuels: oil, gas and coal are the remains of plants and animals from millions of years ago. 86 per cent of world energy comes from fossil fuels.

Nuclear: because nuclear reactors are powered by a non-renewable element, uranium, nuclear is classed as non-renewable energy.

Sugarcane, grown as biofuel in Brazil

Advantages and disadvantages

- **Biomass:** carbon neutral: the CO_2 plants release when burnt = the CO_2 absorbed when growing. However, deforestation is linked to clearing land for biofuel crops.
- **Wind:** no pollution or carbon emissions, but people object to their impact on the landscape.
- **Hydro:** reservoirs mean water can be released to generate power when needed – but whole valleys are flooded which is a disadvantage.
- **Solar:** colossal free energy source – except at night! Clouds also reduce power.
- **Tides:** regular, predictable and, like waves, very powerful, but we do not yet have the technology to exploit them effectively.
- **Geothermal:** very efficient source, but potentially, sites could cool down over time, reducing the energy supply.

Fossil fuel extraction: The North Sea

The first oil fields were discovered in the North Sea in the 1960s. By 2010, three-quarters of all the oil in the UK's North Sea fields had been extracted. In that year, 100 million tonnes of oil and gas was extracted.

- 👍 In 2010, North Sea oil companies paid the government £10 million in tax. In 2011, 450000 people worked in the industry.
- 👎 When oil prices fall, as after 2014, North Sea oil is not profitable because it is difficult to drill. In 2015, the industry cost the government £24 million and 12000 jobs were lost as production slowed.

Now try this

Describe and explain how the extraction of a fossil fuel that you have studied has both advantages and disadvantages. **(6 marks)**

Sustainable energy supplies

Moving towards sustainable energy means increasing the amount of energy from renewable sources, and using energy more efficiently to conserve energy and reduce waste.

Energy conservation at home

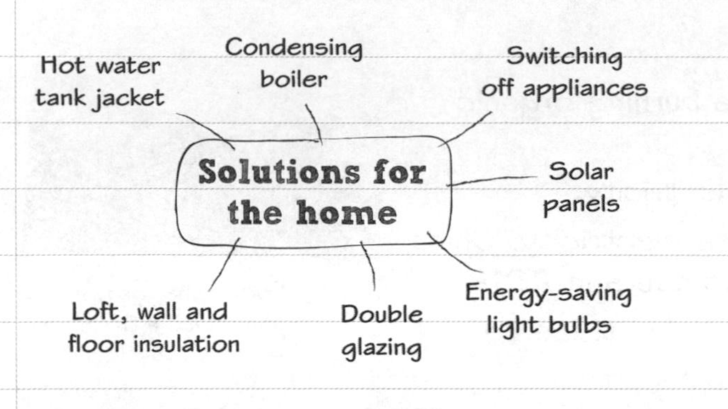

Reducing demand

Governments can aim to reduce energy demand through:

- public education campaigns: for example, to turn down central heating, to walk or cycle instead of driving short distances
- incentives: for example, tax cuts for improving business energy efficiency
- spreading best practice: for example, the best way to replace old-fashioned street lighting with LED lights.

Sustainable workplaces

- Use low-energy lights and hand driers that switch off automatically
- Reduce water wastage with auto on-and-off taps, low-flush toilets
- Reduce printing out with smart scanners that can email documents

Sustainable transport

- Reduced carbon emissions vehicles
- Improve public bus and tram services and extend routes close to workplaces
- Encourage carpooling through social media

Technology and fossil fuels

Carbon capture and storage (CCS) technology cuts carbon emissions from coal-fired power stations by around 90 per cent. It is expensive and reduces the power produced by the coal.

Another (expensive) technology is **integrated gasification combined cycle (IGCC)**, which turns coal into a cleaner energy source called syngas.

Example: Micro-hydro in Bolivia

The NGO Practical Action is developing micro-hydro in Amaguaya, a mountain village of 90 families in Bolivia. The steep mountainsides and fast-running rivers are perfect for micro-hydro. Before, people used candles and kerosene for light. Now the community has a 60 kW plant that provides clean, cheap reliable electricity for schools, a health centre and people's homes.

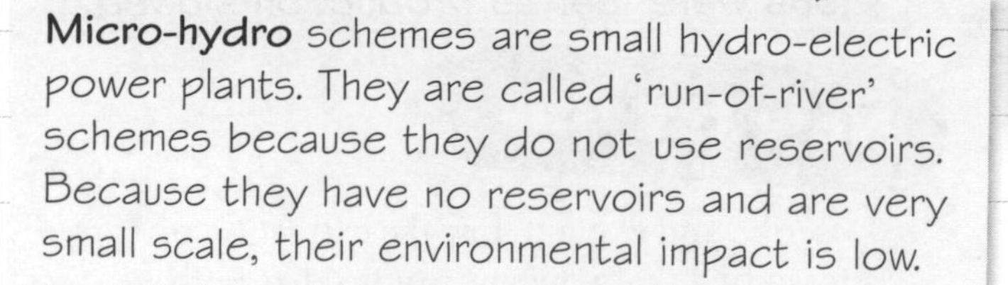

Micro-hydro schemes are small hydro-electric power plants. They are called 'run-of-river' schemes because they do not use reservoirs. Because they have no reservoirs and are very small scale, their environmental impact is low.

Worked example

Complete the following sentences. **(2 marks)**

A person's carbon footprint measures all the *greenhouse gases* they contribute to the environment, measured in kilograms (kg) of *carbon dioxide* per person.

Now try this

Describe and explain how a local renewable energy scheme in an LIC or NEE that you have studied provides sustainable supplies of energy. **(6 marks)**

Working with a resource booklet

Twelve weeks before your Paper 3 exam, you will get a copy of your resource booklet. This is a pack of information about a geographical issue that you will use in your Paper 3 exam.

What is in the resource booklet?

Paper 3 Section A is about applying your geography skills and understanding to questions based on the resources in the booklet.

- There will be lots of different types of resources: for example, maps at different scales, different kinds of graphs and photos, and lots of text information.
- These resources will all be related to the same geographical issue, and will show different aspects of the issue.

You will use these resources in Section A of the Paper 3 exam. You will not know what the questions are until you get into the exam.

There are 12 weeks between getting the booklet and the Paper 3 exam, so you will have plenty of time to get to know the booklet contents really well.

Which topics are compulsory topics?

- Paper 3 Section A can be about any of the compulsory topics in the specification, and the topics are likely to change each year.
- The compulsory topics are shown in the diagram opposite. The issue in the resource booklet will also involve links between topics as well as geographical skills.
- Each topic has a range of different issues associated with it. Your teacher will also help you to identify the sort of issues suggested by the resources selected for the resource booklet.

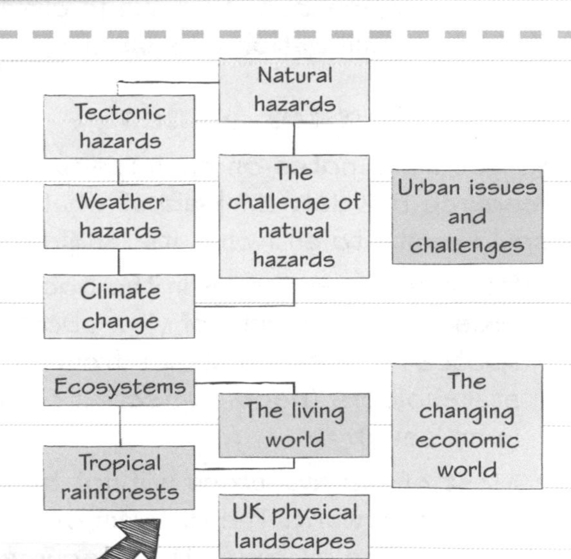

Compulsory topic areas

Making notes on the resources

- Use a highlighter pen to identify key terms, maybe using a different colour for terms that you need to recap.
- Summarise what each resource shows. Consider its limitations, too: what makes it more or less useful.
- Practise answering different kinds of questions from the resources. You can see what sort of questions are asked from looking at sample questions from exam papers and from your teacher.

You cannot take your notes into the exam.

Notes on Figure 5: project file

Project 1: water supply and sanitation
- Funded by aid agency
- Limitations: no info on cost, no info on how many people will be helped

Project 2: water supply, sanitation, housing, fuel
- Joint government/self-help
- Strengths: info on numbers helped
- Limitations: no info on cost

Contemporary geographical issues

Some geography topics are not controversial: for example, different types of river erosion. However, with other topics there is the potential for different points of view. It is that potential for difference that makes a geographical issue. Contemporary means 'happening now' – present-day issues.

The resource booklet and issues

Although you will not know until you get into the Paper 3 exam what issues the questions will focus on, the resource booklet can give you some useful clues. For example:

- if the resources are about urban growth in NEEs and LICs, issues could be about how best to tackle the problems of rapid urban growth
- if the resources in the booklet are about impacts of climate change, issues could be about how best to manage climate change – mitigation or adaptation.

The challenge of natural hazards

Why are the impacts of natural hazards often greater in LICs than in HICs?

Why do people choose to live in areas at risk from a tectonic hazard?

What should be done to reduce the impact of natural hazards?

Is weather becoming more extreme?

What is the evidence for climate change?

What should be done to manage climate change?

Natural hazards: examples of possible issues

Preparing for issues

As you make notes on each resource in the resource booklet, think about what issues it could relate to and what its 'angle' is.

- Sometimes resources in the booklet will present clear points of view about issues: quotes or speech bubbles from different stakeholders (people affected or involved in an issue).
- Most of the resources will not be this straightforward. The diagram opposite would be an example. This resource relates to uneven development – inequalities in wealth. It is presented in a way that maximises the inequality in global income.

The distribution of global income in 2013

Richest

Each band = 20% of the world's population

Poorest

World population	World income
Richest 20%	82.7%
Second 20%	11.7%
Third 20%	2.3%
Fourth 20%	1.9%
Poorest 20%	1.4%

Which issue does this resource relate to?

Issues questions

Not all the questions in Section A of the Paper 3 exam will be about issues, but all the higher mark questions will be about issues.

The exam paper will tell you which parts of the resource booklet to use for each question.

You cannot take your copy of the resource booklet into the exam: you will be given a new copy of the booklet in the exam.

Your own knowledge and understanding

The resource booklet provides a lot of information for you to use in your answers.

However, watch out for gaps in the resources. Use your understanding of geographical issues to look out for limitations. For example, if the booklet lists human causes of climate change and not natural ones, be prepared to fill this gap with your own knowledge.

Evaluating issues

One or more of the questions in Section A of Paper 3 will ask you to make a decision and explain why you have made that decision.

Options

- The resource booklet will provide you with information about the options involved in the decision question.
- You will not know what the issue is that you need to decide on until you are in the exam, but your teacher will help you explore likely options.
- You need to use information from the resource booklet and your own knowledge and understanding to answer the question.

Three projects have been suggested to help the Philippines prepare for future tropical storms. These are described in **Figure 5**.
Which of the three options provides the best protection from future high-intensity tropical storms for the people of the Philippines?
(9 marks plus 3 SPaG marks)

The three options are described in the resource booklet. You will not have the exam question until the exam, but you can practise likely decision-making activities based on the resource booklet before the exam. You need to know the resource booklet well.

Evaluation skills

Evaluation is about weighing up the three options and deciding which one you think is best.

- Decide on at least two advantages and disadvantages (or strengths/weaknesses) for each option.
- Bring in your geographical understanding from the whole course, and use evidence from the booklet to back up your points.
- Discuss the arguments against, as well as the arguments for – explain why you have rejected the other two options.
- You need a balanced argument – considering physical and human factors, economic and environmental factors, and so on, helps with this.

SPaG

SPaG means spelling, punctuation and grammar. In Section A of Paper 3, the decision-making extended prose question will include 3 SPaG marks. A tip for SPaG is to divide your answer into well-punctuated paragraphs. Many student answers 'don't stop for breath', which can make their points hard to understand.

Good news!

There is no single 'right' answer to evaluation questions. Three students could each choose a different option and each still get top marks if they make convincing arguments using the full range of available information together with their own knowledge and understanding, and SPaG!

Now try this

Practise answering the 9-mark (3 SPaG) questions on page 128.

For more on extended prose, see pages 127–129.

Enquiry questions

Section B of Paper 3 is the fieldwork exam. In your exam, you will be asked questions about both of your fieldwork enquiries (human and physical geography), and also some questions where you will need to apply what you learned during your fieldwork to new situations.

Exam questions

Enquiry questions are the kind of questions that can be investigated by fieldwork. They give fieldwork a purpose. You will have put together enquiry questions for your fieldwork.

Exam questions about geographical enquiries focus on how a geographer comes up with a suitable question for a geographical enquiry:

- the factors that need to be included in an enquiry question
- what geographical theory the enquiry is related to
- what types of evidence should be collected (including primary and secondary data)
- the locations for the fieldwork
- risks and how these could be reduced.

Enquiry questions

- An enquiry question often relates to a geographical theory: the sort of theory that can be tested through fieldwork.
- Key questions/hypotheses follow from the enquiry question, and they can be tested.

For example, an enquiry question could be:

- How do different methods of coastal management create benefits and conflicts?

A key question following on from this could be:

- Do people prefer hard engineering to soft engineering in managing coastal processes?

Worked example

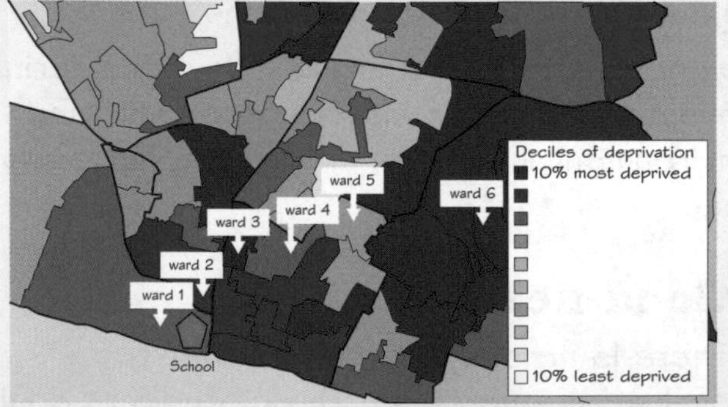

Study this map, which shows levels of multiple deprivation in wards of Brighton in 2010. Students at the school marked on the map decided that ward 6 would not be an appropriate place to carry out fieldwork on variations in quality of life in Brighton's inner city wards. Explain **one** reason why this ward was not appropriate. **(2 marks)**

Ward 6 is a relatively large ward, which would have made carrying out a survey there very time consuming, especially as ward 3, which is smaller, has the same deprivation level.

Geographical examples and theories

You need to be able to identify the key geographical concepts that the investigation is based on.

For example, consider the enquiry question: **How and why do beach profiles vary in [name of coastal location]?**

To be able to evaluate different aspects of this investigation you would need to understand that beach profiles are affected by: wave type; wave frequency; wave direction/longshore drift; local geology; pebble size; beach management strategies (for example, groynes).

Now try this

Describe the location of **one** of your fieldwork enquiries. Explain why it was a good place to investigate your enquiry. **(4 marks)**

Selecting, measuring and recording data

Your fieldwork exam may have questions about the way you went about selecting, measuring and recording data in your own fieldwork, or ask you to apply your knowledge about selecting, measuring and recording data to new situations.

Exam questions

Exam questions about selecting, measuring and recording data focus on **appropriateness**: what is the appropriate type of data to collect for a chosen enquiry, and what are the appropriate ways of collecting it?

- The differences between primary and secondary data: what is each appropriate for?
- Identifying and selecting the appropriate physical and human data to collect for a chosen enquiry.
- Measuring and recording the data using appropriate sampling methods.
- Being able to describe data collection methods and justify the choice of method that were used in an enquiry.

Worked example

As part of an enquiry collecting primary physical geography data, a group of students measured stream depth by stretching a rope across the stream and measuring stream depth using a 30 cm ruler at intervals along the rope.
Suggest **two** ways in which this data collection technique could be adapted to make the sample more reliable. **(2 marks)**

First way: If the students used a tape measure instead of a rope, they could measure depth at set intervals along the tape measure.

Second way: A metre stick would be better than a ruler in case the stream was over 30 cm deep at any point.

Key terms

Primary data: fieldwork data that you collect yourself.

Secondary data: information that has been collected by other people.

Quantitative data: data that can be measured in numbers and written down as numbers.

Qualitative data: data that record qualities of things, such as photos, sketches, surveys about people's views or feelings.

Different kinds of data have different advantages and disadvantages for enquiries.

Worked example

Justify **one** primary data collection method used in relation to the aim(s) of your **human** geography enquiry. **(3 marks)**

Method: Urban land use transect

My enquiry was: How has tourism affected the environment of Ironbridge? The urban land use transect meant I could accurately record all the different shops and other land uses along either side as I walked along the High Street. I also used a category chart to identify what category each shop belonged to in order to help make the data consistent.

Now try this

Study these two images. Which one shows a quantitative fieldwork method and which one shows a qualitative fieldwork method? **(2 marks)**

Measuring water quality by sampling

Collecting views on river flooding by questionnaire

Processing and presenting data

Your fieldwork exam may have questions about the way you processed and presented your data, or ask you to apply your knowledge to new situations.

Exam questions

Exam questions about processing and presenting data focus on **appropriateness**: what is the appropriate way to process and present the fieldwork data for a particular geographical enquiry? You should know about:

- selecting the appropriate presentation method for the enquiry
- using presentation methods accurately
- how to describe and explain different presentation methods.

There are many different ways to process and present data, such as images and other visuals, graphs and maps. For more on some of these methods, see pages 138–140.

Advantages and disadvantages

The exam may ask questions about how the presentation of fieldwork data could be improved, or about alternative methods of presenting data. Use your knowledge of advantages and disadvantages of different types of presentation for this.

Worked example

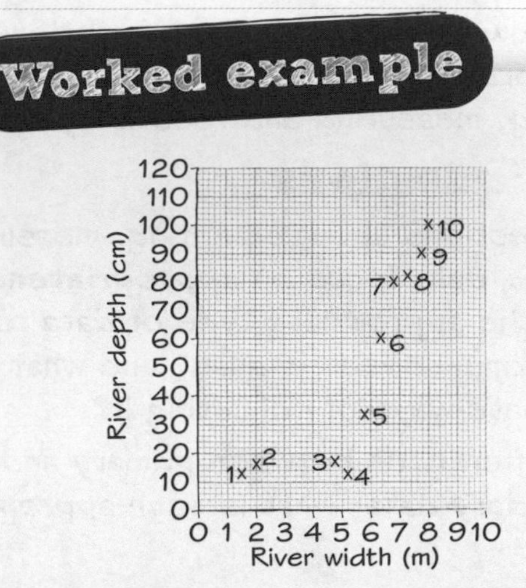

A student investigated the relationship between river depth and river width, and presented the data using a scattergraph.

Explain **one** advantage of using a scattergraph to present these data. **(2 marks)**

An advantage of a scattergraph is that it can be used to identify a relationship between two variables. A line of best fit can be drawn to indicate whether the relationship is positive, negative or whether there is no relationship.

Presentation advantages

This is a list of the top five advantages.

1. Isopleth maps: ideal for showing gradual change over an area.

2. Proportional symbols maps: very accessible and easy to understand.
3. Kite diagrams: show changes over distance (e.g. transect data).

4. Dot maps: give a clear indication of differences in density for a geographic distribution (e.g. for tourist signage).
5. Flow maps: show direction and volume of movement (e.g. for vehicle counts).

Presentation disadvantages

This is a list of the top five disadvantages.

1. Scattergraphs: can only show relationships between two variables.

2. Pie charts: lots of small segments make the chart difficult to interpret.
3. Choropleth maps: hide variations within areas.
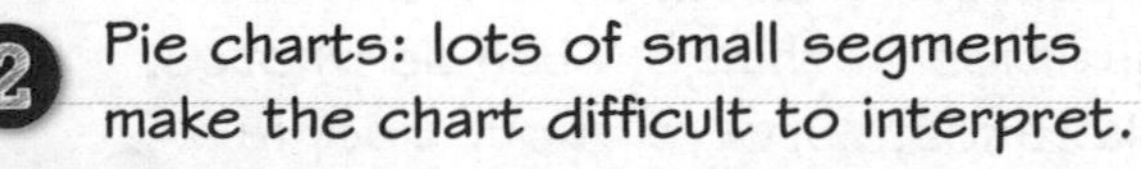
4. Triangular graphs: data must be in %.
5. Bar graphs: do not show relationships between categories.

Now try this

Describe the data presentation method(s) you used for one of your geographical enquiries and explain why it was appropriate. **(4 marks)**

Analysing data and reaching conclusions

Exam questions

Exam questions about analysis and conclusions focus on **appropriateness** and **evidence**: What is the appropriate way to analyse data for a particular geographical enquiry? Are conclusions are based on evidence?

You should know about:

- describing, analysing and explaining the results of fieldwork data
- making links between different sets of data
- using the appropriate statistical technique to analyse data
- identifying anomalies (things that do not fit a pattern) in fieldwork data.

Analysing data

Key steps for successful data analysis:

1. **Describe** what you see:
   - What are the overall patterns or main features?
   - Are any figures or features in groups?
   - What about anomalies or exceptions?
2. Use **evidence** – precise numbers or facts from the data – in your analysis.
3. Give **reasons** for the patterns you see.
4. Link these reasons to **geographical** concepts/theories if you can.

Worked example

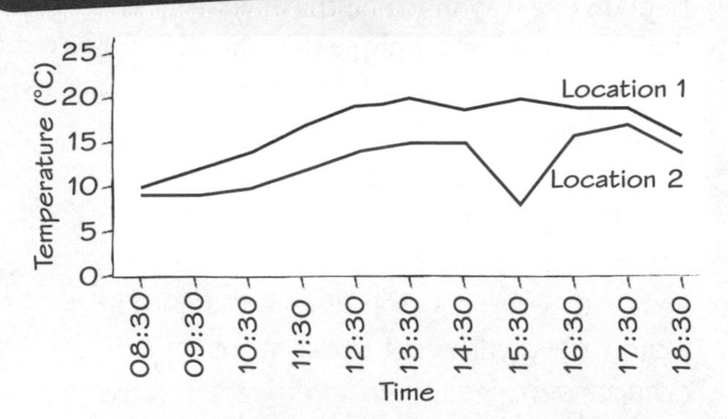

As part of an enquiry collecting primary physical geography data, a student measured the air temperature at two locations every hour from 8.30 to 18:30 on 18 July 2016.

Describe the results of this fieldwork investigation. **(2 marks)**

Location 1 was significantly warmer (between 2° and 5°) than Location 2, with a larger range (10°, compared to 8°). Location 2: the data includes an anomaly at 15:30, which suggests a measurement error.

This is a good answer because it has used specific information from the resource to enhance its description with accurate figures. The answer also uses a precise term, anomaly, to describe the data in the question.

Conclusions and summaries

In your conclusion you should go back to your key question or hypothesis, and use evidence from your investigation to answer it.

In the exam you may be asked to reflect on aspects of your investigation. You will need to either 'assess' or 'evaluate'.

- To **assess** you need to think about all the factors and identify the most important ones.
- To **evaluate** you need to weigh up the value or success of something, and come to a conclusion.

Now try this

The table shows beach profile data for a stretch of coastline. Calculate the mean beach profile angle. **(1 mark)**

Profile number	1	2	3	4	5	6
Profile angle in degrees	3.0	2.8	3.9	2.0	1.0	2.6

Evaluating geographical enquiries

Your fieldwork exam may have questions about the way you evaluated your geographical enquiry, or ask you to apply your evaluation skills to new situations.

Exam questions

Exam questions about evaluating your enquiry (or evaluating fieldwork from an unfamiliar context/new situation) will focus on how things could have been improved, or the problems you experienced and how they could have been avoided or solved. They may also focus on **reliability**. You should know about:

- identifying problems with data collection methods
- identifying limitations of data collected
- making suggestions for other data that might have been useful for the enquiry
- assessing whether the conclusions that were reached were reliable.

Conclusions might not be reliable if they are based on data that was limited or unsuitable in some way, or on data analysis techniques that were not appropriate.

Worked example

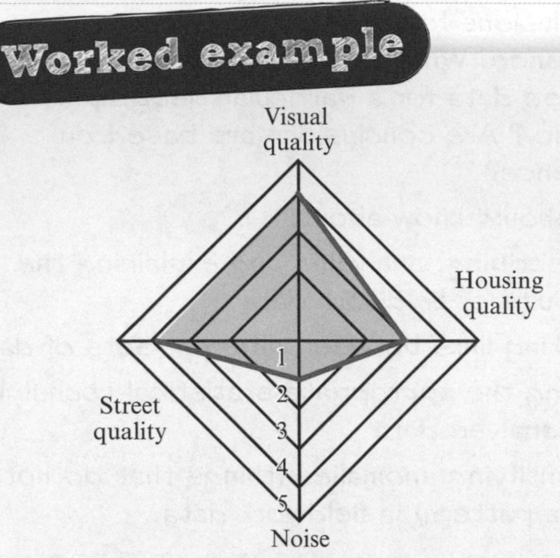

A student investigated environmental quality in different urban locations. She presented her data using a radar graph, one graph for each location. Explain **one** way in which the effectiveness of the presentation technique that she used for her enquiry could have been improved. **(4 marks)**

An advantage of a radar graph is that it can display data on several different variables, so it is a good way to compare the different characteristics of an area. However, a combined radar graph would have allowed for a more effective comparison between her five surveyed locations: plotting each one separately was time consuming and not as easy to use as one combined graph.

Five evaluation tips

1. Evaluating the limitations of your enquiry is not the same as saying your enquiry went wrong or was no good. Every enquiry has limitations and ways in which it could be improved.
2. A good place to start is your own experience. What problems did you encounter on your fieldwork trips? Then think of what you would do differently if you did that part of the fieldwork again.
3. Data collection methods all have advantages and limitations, for example, sample size. You can use your knowledge of these when evaluating your enquiry.
4. Data presentation methods, including GIS, have advantages and limitations, too. These are another good way to evaluate your enquiry.
5. The limitations of primary and secondary data is another good area to focus on. It is often difficult to collect enough primary data for statistically reliable results, for example.

Now try this

Identify **one** limitation of **quantitative** data and **one** limitation of **qualitative** data. **(2 marks)**

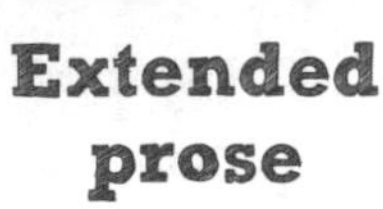

6-mark questions

Extended prose questions are questions where you need to write a longer answer. In the exam, some extended prose questions will be worth 6 marks.

What do 6-mark questions look like?

'Weather is becoming more extreme in the UK.' Use evidence to support this statement. **(6 marks)**

Use Figure 1 and your own knowledge to **explain** how problems with access to water can affect people's standards of living. **(6 marks)**

Most 6-mark questions come with a resource. How you select and use information and pick out the main trends and patterns will be important.

Discuss the impacts of international migration on city growth. Use Figure 1 and a case study of a major city in the UK. **(6 marks)**

Describe and **explain** the economic problems caused by traffic congestion. **(6 marks)**

Explain the costs and benefits of a 'do-nothing' approach to coastal management. **(6 marks)**

Explain the processes involved in the formation of the landforms shown in Figure 1. **(6 marks)**

'Hot deserts have excellent potential for energy development.' Use evidence from your case study to support this statement. **(6 marks)**

To what extent do hard-engineering strategies offer the best form of defence against river flooding? **(6 marks)**

6-mark questions use command words like the bold ones in these examples. Identify the command word or phrase before you start writing to help focus your answers.

How are 6-mark questions marked?

Three levels are used:

- **Basic** (little supporting detail; not straightforward to read or understand) – these answers are worth 1–2 marks
- **Clear** (straightforward to read and understand; not enough supporting detail) – these answers get 3–4 marks
- **Detailed** (balanced; well-developed ideas; a thorough understanding of the topic) – these get 5–6 marks

Answers with no relevant content get 0 marks.

Assessment objectives (AOs)

6-mark questions target AO1 (2 marks) and AO2 (4 marks).

- ✓ **AO1** is about showing your **knowledge**: for example, your knowledge about the processes of river erosion and deposition.
- ✓ **AO2** is about showing your geographical **understanding**: for example, your understanding of how erosion and deposition combine to form meanders.

Where do 6-mark questions appear?

6-mark questions can appear in any paper or section of the exam. They are the second most valuable exam question type, so leave time to answer them clearly and in detail.

Revising for 6-mark questions

For each topic, look out for problems, impacts, processes, costs and benefits, and issues where there is more than one point of view. These will show your understanding.

Make sure you learn your case study, named examples and examples so you can provide details and evidence to show your knowledge.

Now try this

'6-mark questions take many different forms.' Use evidence to support this statement. **(6 marks)**

Had a look ☐ Nearly there ☐ Nailed it! ☐

9-mark questions

The longest answers you write will be for the 9-mark extended prose questions. Some of these will also have marks available for spelling, punctuation and grammar (SPaG).

What do 9-mark questions look like?

Assess the extent to which the economic impacts of extreme weather effects are more significant than the social or environmental impacts. Use the text extract above and an example you have studied. **(9 marks + 3 SPaG)**

Explain the advantages and disadvantages of TNCs (transnational corporations) to one LIC or NEE host country that you have studied. **(9 marks)**

'Economic development only has positive effects on quality of life for people in LICs and NEEs.'
Do you agree with this statement?
Yes ☐ No ☐
Justify your decision. **(9 marks)**

Using Figure 1 and your own knowledge, **explain** how different landforms may be created by the transport and deposition of sediment along the course of a river. **(9 marks)**

To what extent does a cold environment you have studied provide both opportunities and challenges for development? **(9 marks)**

Evaluate the effectiveness of an urban regeneration project you have studied. **(9 marks + 3 SPaG)**

Assess the opportunities for development of a cold environment you have studied. **(9 marks)**

How are 9-mark questions marked?

Extended writing is marked by level three.

- **Basic** – these answers are worth 1–3 marks
- **Clear** – these answers get 4–6 marks
- **Detailed** – these get 7–9 marks

Answers with no relevant content get 0 marks.

A detailed answer is balanced, has well-developed ideas and good understanding of the topic.

SpaG

SPaG stands for spelling, punctuation and grammar, and it also includes using specialist vocabulary. **SPaG is only assessed in 9-mark questions**, making a further 3 marks available for some questions.

Where do 9-mark questions appear?

- **Paper 1**
  There will be a 9-mark question in both **Section A** and **Section B** (not Section C). **One** will also be assessed for SPaG (+ 3 marks).
- **Paper 2**
  This is the same as for Paper 1.
- **Paper 3**
  Sections A and B will each have a 9-mark question assessed for SPaG (+ 3 marks).

Assessment objectives (AOs)

9-mark questions target AO1 (3 marks), AO2 (3 marks) and AO3 (3 marks).

- **AO1** is about showing your **knowledge**.
- **AO2** is about showing your geographical **understanding**.
- **AO3** is about **applying** your knowledge and understanding to do highly-skilled things like evaluating and making judgements about geographical issues.

Now try this

To what extent are 9-mark questions the most challenging questions on the AQA geography exam paper? **(9 marks)**

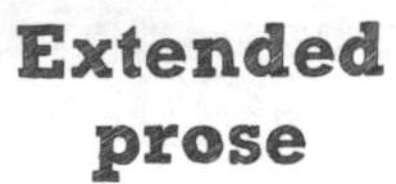

Paper 3 has two sections – **A** and **B** – and each ends with a 9-mark question (plus SPaG). These are a bit different from the extended prose questions in Papers 1 and 2.

Paper 3, Section A

Section A is called **Issue evaluation**. This is where you will work with the resources booklet.

- There are two **6-mark** extended prose questions in Section A, which work in the same way as those in Papers 1 and 2.
- There will also be **a decision-making question** – the **9-marks (+ SpaG)** question.

For more on working with the resource booklet, see page 119.

Paper 3, Section B

Section B is where you answer questions about your own **fieldwork** (your two geographical enquiries) and other fieldwork scenarios.

The extended prose questions in Section B focus entirely on assessment objective 3. AO3 is all about **applying** what you know – so you will need to **assess** and **evaluate your fieldwork**.

Decision-making

Your resources booklet holds information on **three** different scenarios or options. So, there might be information on three rainforest deforestation projects in Madagascar:

- ✓ Project 1: a government project to set up a nature reserve where no use of the forest at all was permitted except scientific research
- ✓ Project 2: a community-led sustainable forestry project, in which farmers received money to replant land with trees
- ✓ Project 3: a plan to involve the local community in rainforest ecotourism

The final 9-mark (plus SPaG) question asks you to make a decision about the three options.

Which of the three projects do you think will be most successful in reducing the rate of deforestation in this area of Madagascar?
Use evidence from the resource booklet and your own understanding to explain why you have reached this decision. **(9 marks + 3 SPaG)**

The best answers use a **wide range** of information from the booklet to evaluate each option.

How you analyse the options and how you justify your final choice is important.

Assessing and evaluating

Geographical enquiries involve lots of decisions and judgements at every stage, such as:

- What methods of gathering data are most appropriate for my enquiry?
- What is the most appropriate way of presenting my data?
- How reliable are my conclusions?
- How could I have improved different parts of my geographical enquiry?

The extended prose questions ask you about the strengths and weaknesses of your enquiry decisions and judgements of your results and your conclusions. The questions will often use command words like:

- **Assess** how effective ...
- **Evaluate** the effectiveness of ...
- For one of your geographical enquiries, **to what extent** were X helpful in ...

It is important to recognise the way these questions are asking you to weigh up strengths and weaknesses of your enquiry.

Now try this

Match the following command terms to their correct definition: **A** Assess, **B** Evaluate, **C** Explain
1 Judge from available evidence **2** Make an informed judgement **3** Set out purposes or reasons **(3 marks)**

Atlas and map skills

Throughout your GCSE course you will have developed a wide range of different geographical skills – working with maps, graphs, statistics and data. In your exam, questions involving skills can come up at any point on any Paper, so it is definitely worth revising the following pages.

Atlas maps

One of the most common types of map shows **distribution** – for example, the distribution of vegetation types, such as tropical rainforests.

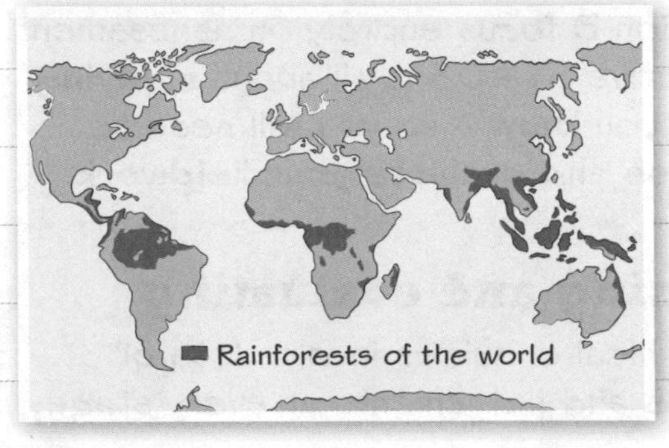

Atlases also contain maps which show:

- **climate zones** and global variations in **precipitation** and **rainfall**
- country boundaries (political maps)
- **height** and **shape** of the land **(relief)**
- population distribution (how people are spread within a region or country).

You may be required to use a combination of different types of information to answer a question.

Describing patterns

Ways of describing a **distribution, pattern** or **trend**: linear, dispersed, spaced, clustered, dense, uneven, scattered, even, sparse, irregular

You can also use the letters GSE to help structure your descriptions of patterns:

G – General overall trend or pattern

S – Specific examples that illustrate the trend or pattern

E – Exception: note any anomalies that do not fit in with the general pattern or trend.

Don't fall into the trap of **explaining** why the pattern happens, unless asked to. Underline the command word in the question to help you focus your answer.

Worked example

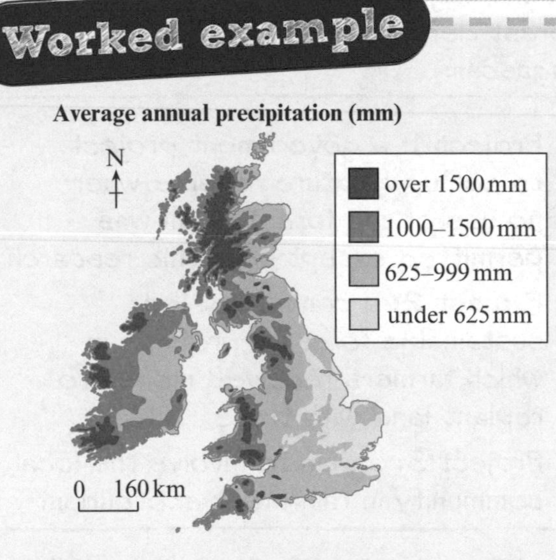

Analyse this figure, which shows the distribution of rainfall in the UK.

Describe the distribution of rainfall shown by the map. **(2 marks)**

Some parts of the UK, in the west and north, receive over 1500 mm of rainfall annually. The east of the UK is drier, with some parts of eastern England receiving less than 625 mm per year on average.

Now try this

Analyse the map above, which shows the global distribution of tropical rainforest.

Suggest **two** reasons for the distribution shown by the map. **(4 marks)**

Types of map and scale

You need to be able to recognise and describe the distribution and patterns shown by the types of maps found in atlases, and deal with maps at a range of scales.

Satellite images and maps

Political maps which show the outline of countries

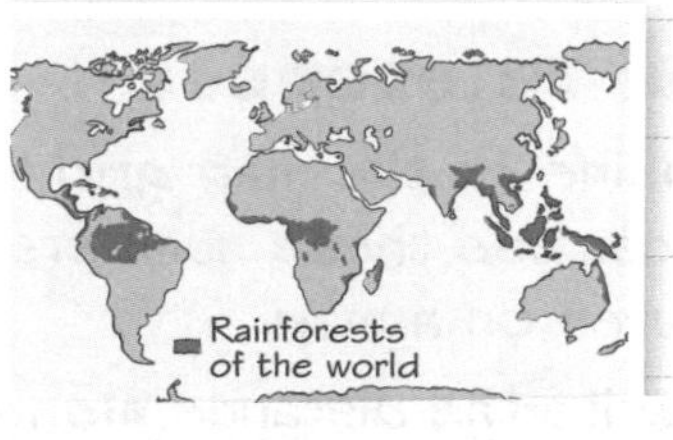

Maps which show the **distribution** of vegetation type, e.g. location of tropical forests

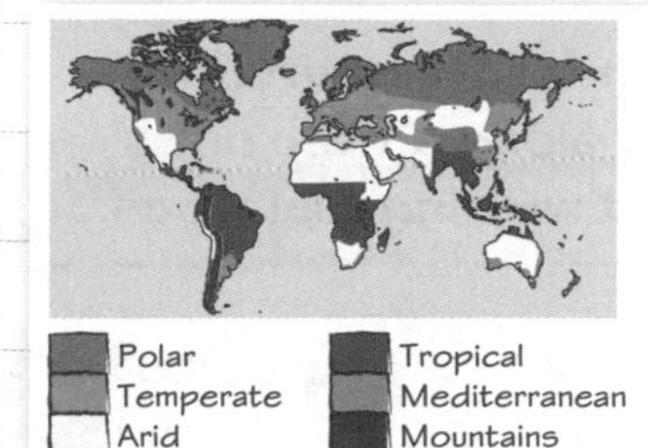

Climate zones which reflect global variations in precipitation and temperature

Relief map of China

level 1: 0–350 m
level 2: 351–1370 m
level 3: 1371–2500 m
level 4: 2501–5490 m

Relief maps showing the **height** and **shape** of the land

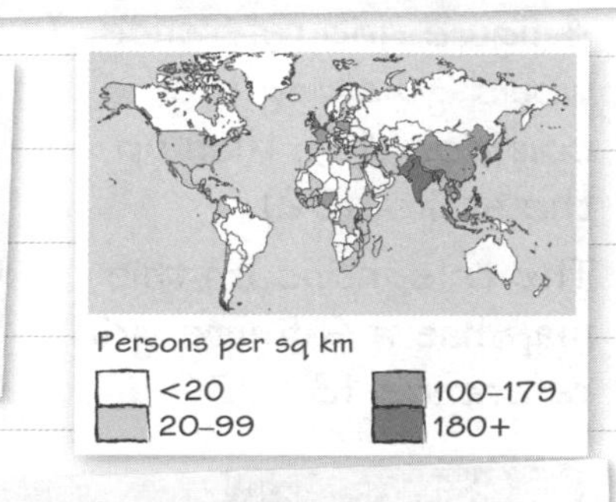

Population distribution maps which show how **spread** out people are within an area

What is scale?

A map's scale tells you how much smaller the area shown on the map is compared to the area in real life.

- For OS maps at 1:25 000 scale, 1 cm on the map represents 25 000 cm in real life (250 m).
- For OS maps at 1:50 000 scale, 1 cm on the map represents 50 000 cm in real life (500 m).

Large-scale maps show a smaller area in more detail.

Small-scale maps show a larger area in less detail.

Worked example

Study this 1:25 000 Ordnance Survey map extract of Bolton Abbey.
How far is it from the car park at Bolton Bridge along the B6160 to the car park at Bolton Abbey? **(1 mark)**

☐ **A** 0.5 km
☒ **B** 1 km
☐ **C** 10 km
☐ **D** 10.5 km

Now try this

The 1:25 000 map extract above of Bolton Abbey shows part of the River Wharfe in North Yorkshire. Which of the following is the best description of the River Wharfe? **(1 mark)**

☐ **A** A lowland river on a wide flood plain
☐ **B** A lowland river entering an estuary
☐ **C** A fast-flowing mountain stream
☐ **D** A river flowing through a gorge

Grid references and distances

Grid references are used to locate geographical features on an OS map. You need to know how to use grid references accurately.

Grid references

Each line on the map grid has a number. You can use these numbers to locate the features on a map.

You write the distance **along** (from the horizontal **easting** line) before the distance **up** (from the vertical **northing** line). For example, the shaded grid square has a 4-figure grid reference. To find this, you would go **along** the corridor, (13), then **up** the stairs, (02).

The telephone on this map has a 6-figure grid reference 138026.

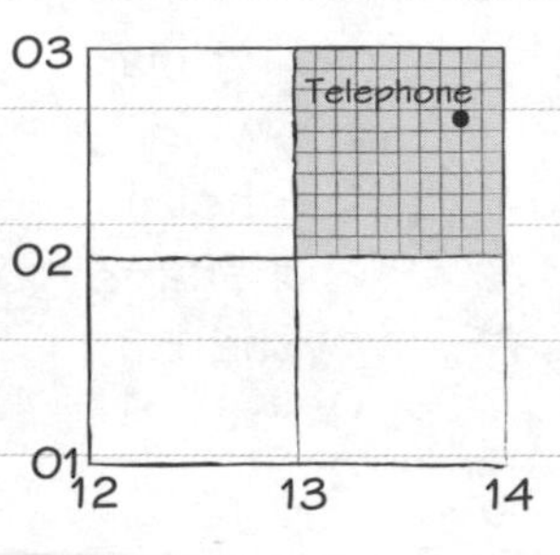

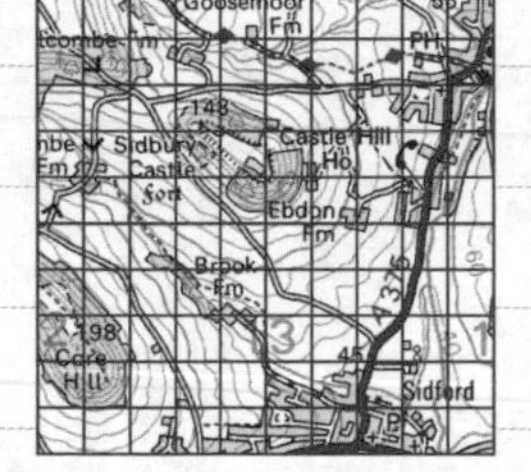

To write a 6-figure grid reference you have to mentally divide each grid square into 10 parts.

Measuring distance

You will need to work out **distances**. There are two types:

1. Distances from one point to another in a **linear** fashion are called **straight line** distances – sometimes called 'as the crow flies'.
2. Distances which follow a **curved** pattern, usually along a river or road, are called **winding** distances.

For the exam you will need a ruler to measure **straight line** distances and a 10-cm-long piece of string for **winding** distances. The string should be provided for you if it is needed in the exam, but bring your own piece of string just in case. Remember, you will be required to convert the distance measured using the ruler or string into kilometres (or metres if more appropriate) – use the scale line on the OS map to help you.

Worked example

Look at the OS map extract.

1 What is the name of the hill in grid square 1292? **(1 mark)**

Bald Hill

2 What is the 6-figure grid reference of the telephone in 1391? **(1 mark)**

137912

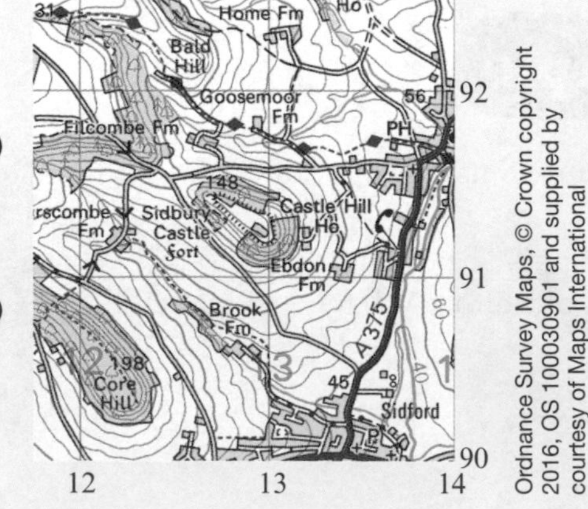

Watch out! A pointer line from the telephone symbol (grid square 1391) shows the **actual** position of the telephone on the main road.

Now try this

There are two scales of Ordnance Survey maps that you will have used in your GCSE course: 1:50 000 and 1:25 000.

If you compared an area on a 1:50 000 scale map with the same area on a 1:25 000 scale map, which would show a higher level of detail? **(1 mark)**

Cross sections and relief

A **cross section** is a visual representation of the landscape from an OS map. You may be asked to draw, label or annotate one, or comment on how you would complete it.

Drawing a cross section

1. Place a strip of paper along the given transect line.
2. Mark off the points where the major (brown) contour lines meet the transect line.
3. Mark the location of other features such as rivers, roads or high points.
4. Draw a line on the grid paper to be the x-axis of your cross section. Line the strip of paper up with this x-axis.
5. Mark off the height of each contour line using a neat cross. Join up the crosses with a ruler and a sharp pencil.

Slopes

The closer the contours, the steeper the slope!

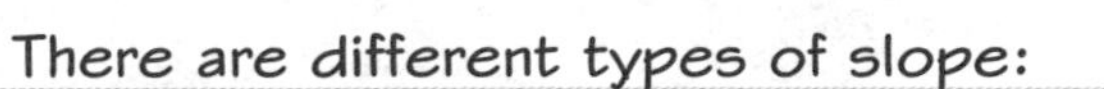

There are different types of slope:

- **concave** slope
- **convex** slope

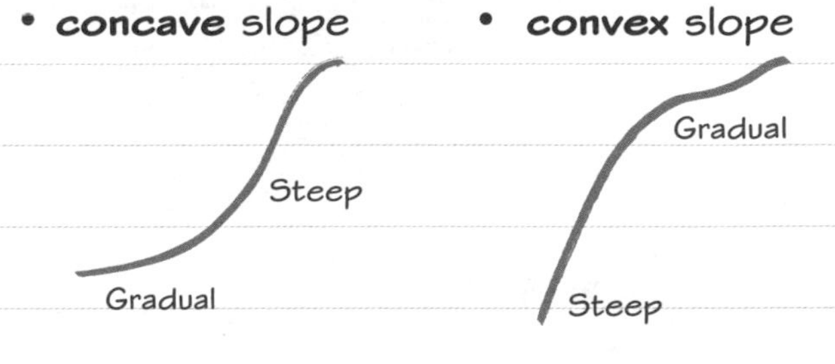

Worked example

Look at the map extract on the right.

Put a cross in the box below to describe the shape of the land from A to B. **(1 mark)**

☐ **A** The land rises more gently towards the west.
☒ **B** The land at B is the highest.
☐ **C** The land at A is the highest.
☐ **D** The river goes through coniferous forest.

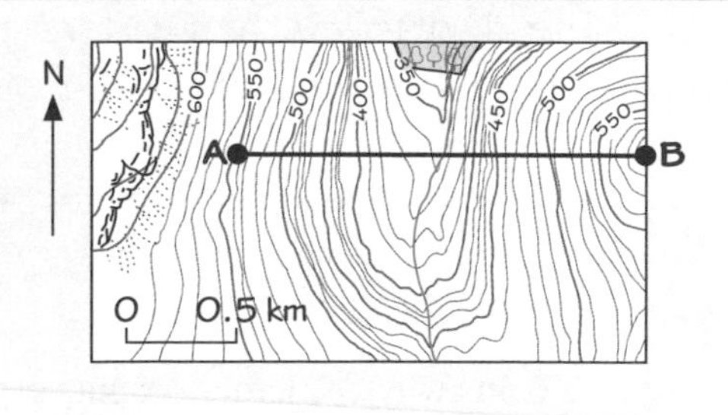

In multiple choice questions look at the number of marks as this is likely to tell you the number of answers required.

Now try this

1 Look at the map extract in the Worked example. Create a cross section of line A–B. **(4 marks)**
2 Explain how spot heights are used on OS maps. **(2 marks)**

Physical and human patterns

You may be asked to use photos, maps or sketches to describe or explain **physical** and **human** patterns.

Describing patterns

You should learn to **describe** and/or **explain** the **distribution** and **pattern** of **physical features** (rivers and coastlines) and **human features** (settlements and roads).

Use the same technique you would use to describe any other type of pattern.

- ✓ You can use maps, photos and sketches to describe an area (e.g. rivers and coastlines).
- ✓ You can describe the site of a settlement using a map (e.g. settlements and roads).
- ✓ A photo or sketch can provide more detail about the **function** of the settlement.

For more about atlas and map skills, see page 130.

Worked example

Describe the section of the River Browney and its valley shown on the map extract. Use map evidence in your answer. **(3 marks)**

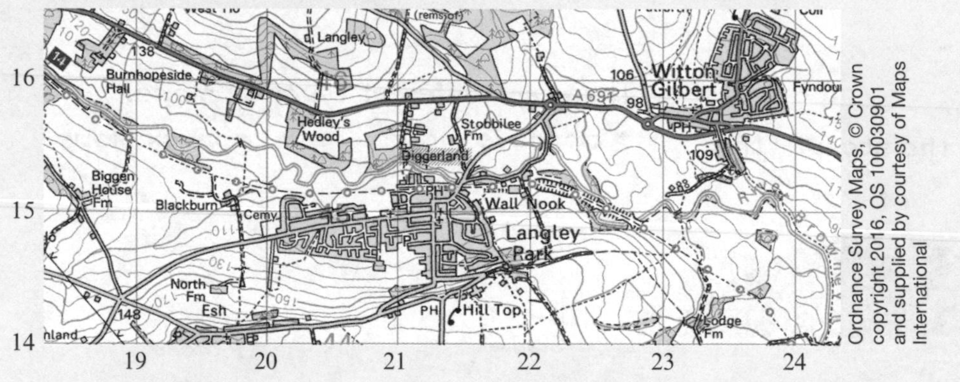

Make sure you use a good range of descriptive comments and map evidence.

This section of the River Browney shows the river moving from a V-shaped valley in the north-west of the map extract into a wider valley in the south-east. The V-shaped valley is only around 100 m wide at river level, with sides sloping up quite steeply to elevations of around 160 m. The river valley then widens to approximately 500 m, and is flat. There are a number of river meanders as the river valley widens, especially at grid reference 205153 and 233152.

Now try this

Suggest reasons why Settlement Y has grown. **(3 marks)**

Remember to focus on the **human** and **physical** reasons for the growth of the settlement.

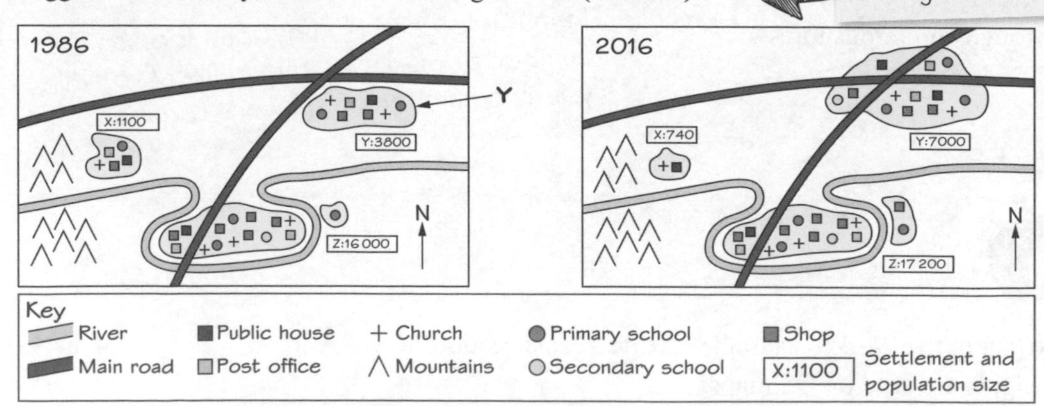

Human activity and OS maps

You need to be able to recognise different types of human activity on an OS map. You may be tested on these.

OS maps may show evidence of different types of human activity.

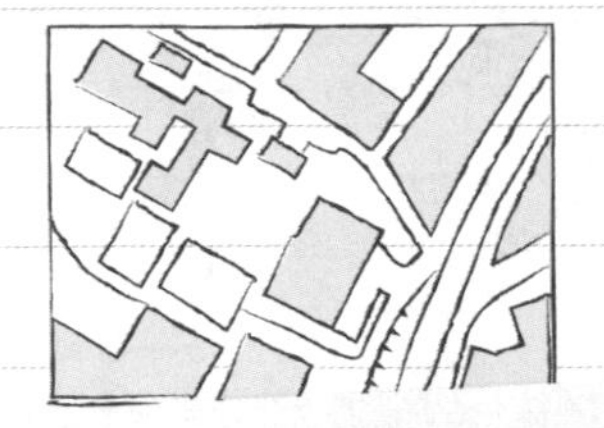

Industrial (e.g. factories and industrial estates)

Residential (e.g. houses and flats)

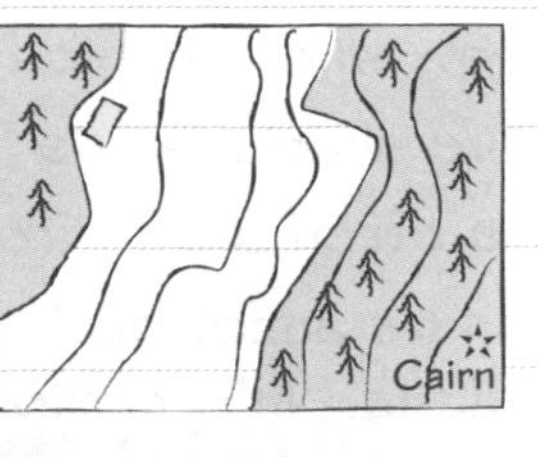

Rural (e.g. forestry and agriculture)

Worked example

Look at the map extract of Ross-on-Wye.
Identify **two** pieces of map evidence that show non-residential activity by humans. **(2 marks)**

At 620228 there is a farm and at 586233 there is an industrial works.

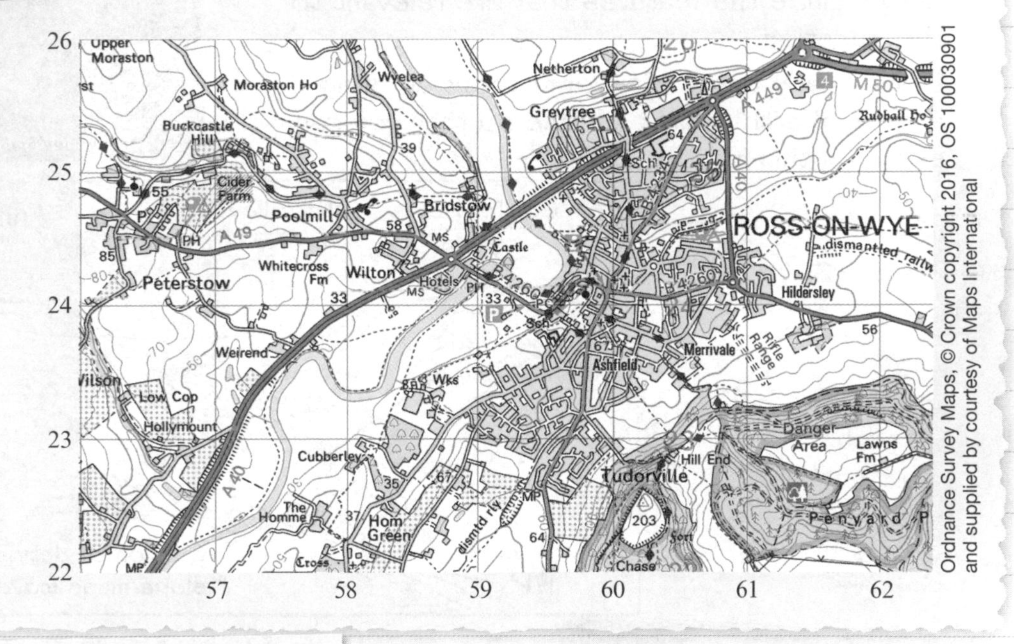

You need to name **two** different types of evidence you can see on the map. You can give a grid reference for each one to show that you know where they are.

Now try this

Study the map extract of Ross-on-Wye on this page.
Identify **two** pieces of map evidence for tourism in Ross-on-Wye and its immediate surroundings. Identify your evidence with 6-figure grid references. **(2 marks)**

Had a look ☐ Nearly there ☐ Nailed it! ☐

Sketch maps and annotations

Drawing sketch maps

Sketch maps can be drawn using information from a map or photograph, or drawn in the field. They:

- ✓ show where basic features are located
- ✓ have simple labels
- ✓ are often drawn from an **aerial** viewpoint
- ✓ can be annotated to add more explanation or detailed information.

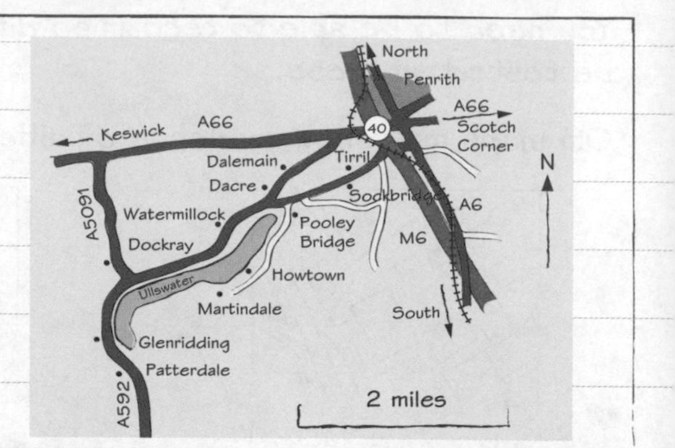

Sketching, labelling and annotating

- Photos and sketches are labelled and annotated in the same way.
- Only include the features that are relevant to the question.
- Draw clearly but don't worry about creating a work of art! Include a frame so that you can sketch within it.

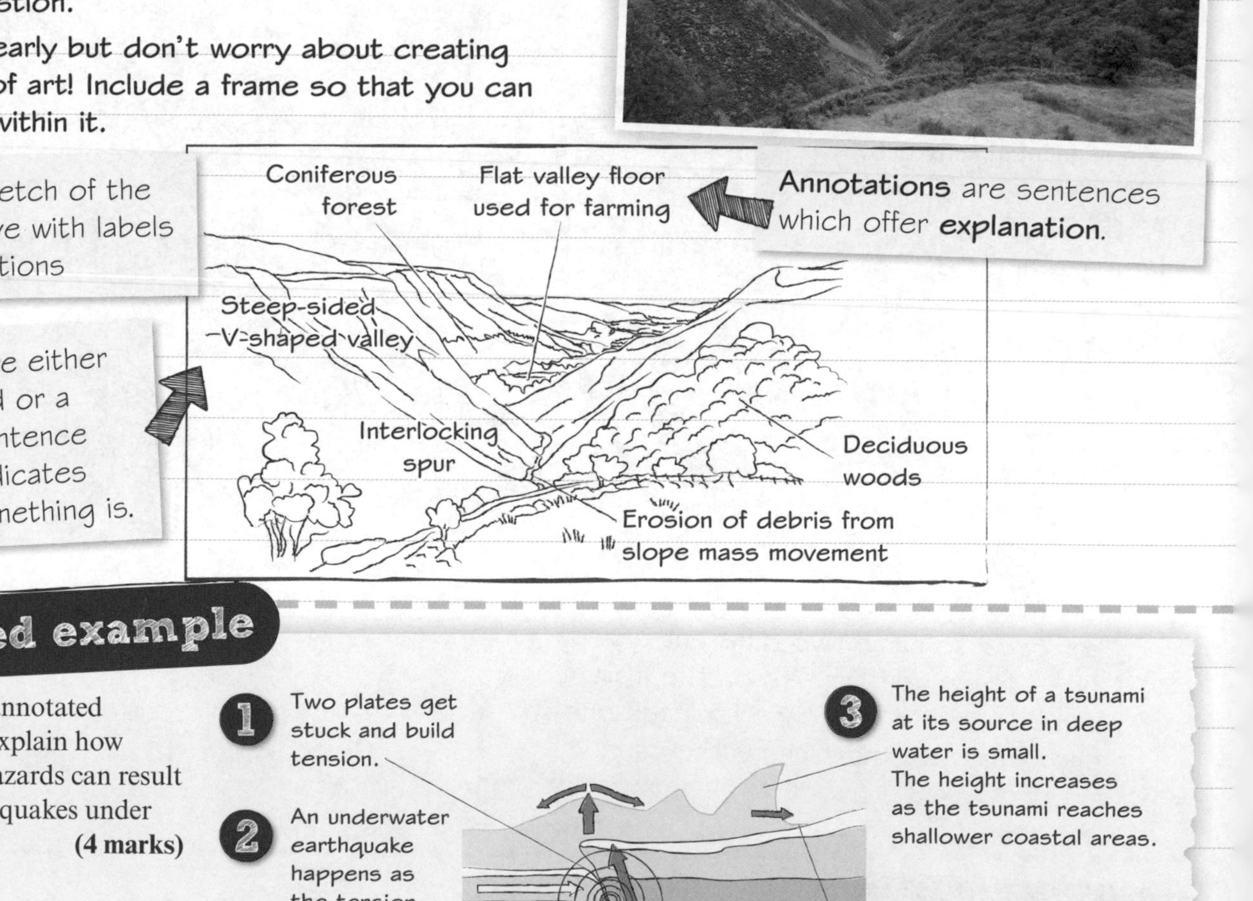

This is a sketch of the photo above with labels and annotations

Annotations are sentences which offer **explanation**.

Labels are either one word or a short sentence which indicates what something is.

Worked example

Using an annotated diagram, explain how tsunami hazards can result from earthquakes under the sea. **(4 marks)**

1. Two plates get stuck and build tension.
2. An underwater earthquake happens as the tension breaks, snapping one plate edge upwards.
3. The height of a tsunami at its source in deep water is small. The height increases as the tsunami reaches shallower coastal areas.
4. The swell reaches the coast and travels very quickly inland.

Now try this

Explain **one** advantage of using annotated diagrams to explain the impact of changes in river or coastal processes. **(2 marks)**

Using and interpreting images

You should be able to respond to and interpret ground and aerial photographs and satellite images.

Different kinds of images

Ground-level photograph: shows lots of foreground detail. Use foreground and background to describe where things are in this type of photo.

Oblique aerial photograph: shows more of the area than a ground-level photo, and features are easier to identify than a vertical photo. But it is hard to judge scale for background features.

Satellite image: measures differences in energy radiated by different surfaces. False colour images convert this data into colours we can recognise. True colour images show us what the satellite sees, for example, vegetation shows up red.

Vertical aerial photograph: these have a plan view, like maps. But details can be hard to identify.

Worked example

Study this aerial photo of a solar farm. Suggest **two** environmental issues that might be associated with this development. **(2 marks)**

The solar farm covers a large area of what was probably previously farmland. This could mean a loss of habitat for plants and animals.

The solar panels are reflecting light. This could be annoying for local residents.

The Five Ws

When working with photos, be sure to remember the Five Ws.

- ✓ **What** does the photo show?
- ✓ **Why** was it taken?
- ✓ **Who** are the people in it (if there are people in the photo)?
- ✓ **Where** was it taken?
- ✓ **When** was it taken (to indicate how long ago it was taken, what time of day, etc.)?

Now try this

Describe **two** ways in which satellite photos like the one shown above can be used to monitor and predict tropical storms. **(2 marks)**

Graphical skills 1

In your exam you may be asked about different types of graphs and charts: to interpret the data they provide and also possibly about when it is appropriate to use a particular chart or graph.

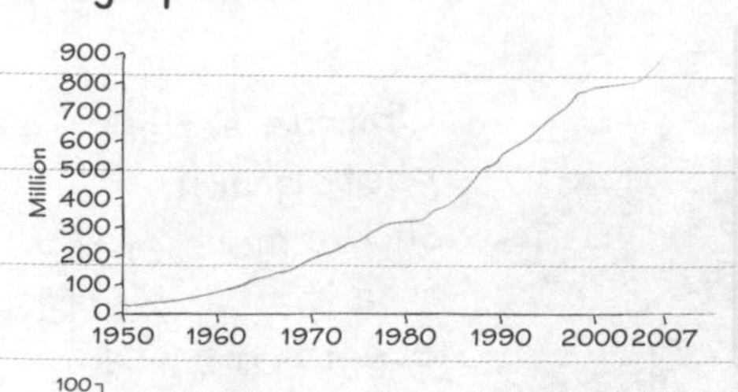

Line charts are used to plot continuous data. They are often used to show how something varies over time. Make sure you plot the points accurately and join the points with a continuous line.

Bar charts are used to plot discontinuous data. Make sure you draw bar charts with a ruler to keep the lines straight.

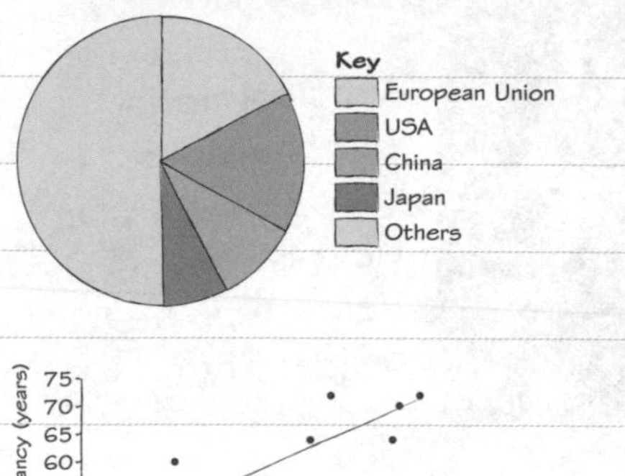

Pie charts show proportions. They are easy to read and fairly simple to put together. Data need to be converted into percentages first and then into proportions of 360° – the whole pie.

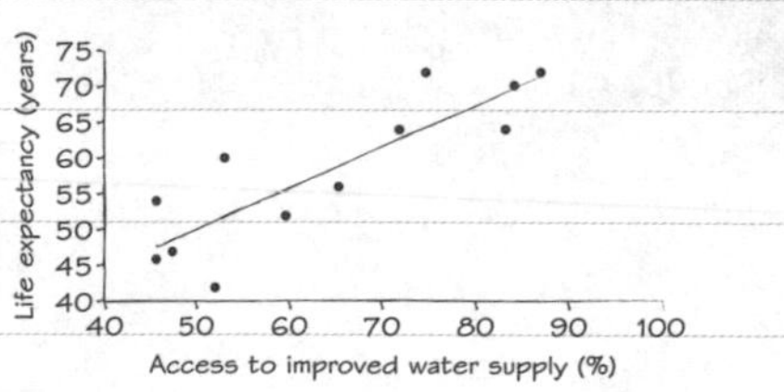

Scatter plots show the relationship between two sets of figures. It is the pattern the points make that is important, so don't join up the points. If the line of best fit slopes downwards it is a negative correlation; if it is upwards it is a positive correlation. Some scattergraphs do not show any relationship.

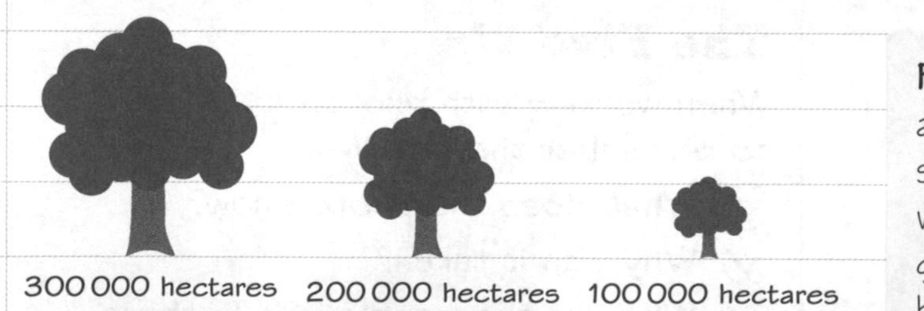

Pictograms represent data using appropriate symbols that are drawn to scale. They present data in a very clear way. However, detailed information can get lost. A key explains the relationship between the data and the pictogram.

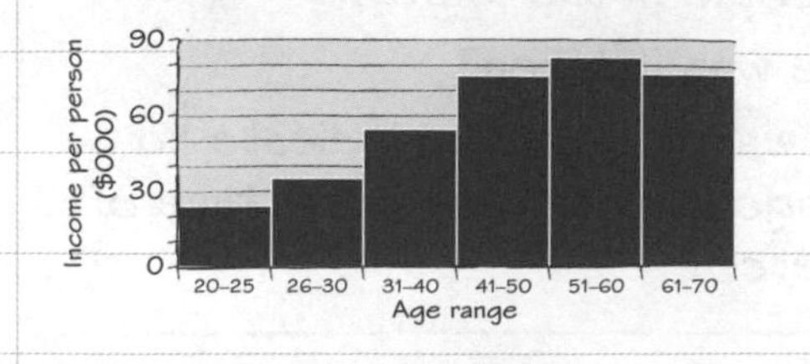

Histograms are used for continuous data. There are no gaps between the bars. The bars should be the same width for each category: this is called equal class intervals. The same colour is used for each bar because the data are continuous.

Now try this

What kind of chart or graph would you use to illustrate the following sets of data?

(a) Population growth in China from 1950 to 2010. **(1 mark)**

(b) The relationship between the size of settlements and the number of services in each. **(1 mark)**

(c) The proportion of people from different ethnic groups living in an inner city area. **(1 mark)**

Graphical skills 2

You should also know how to interpret population pyramids, choropleth maps and flow-line maps.

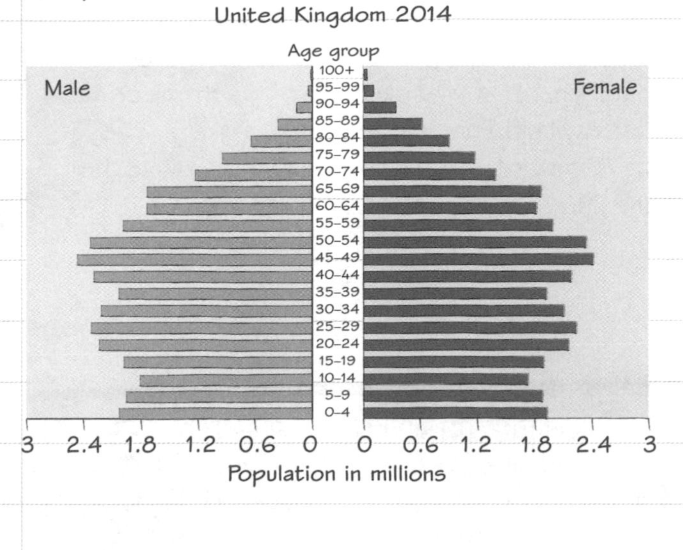

Population pyramids

Population pyramids are graphs that display data about the age structure of a population.

- The data are split into males and females.
- The data are presented in age groups.

This is how to interpret population pyramids:

- Developing countries often show a clear pyramid shape: large numbers of children, lower life expectancy.
- Sides get straighter with development as birth rate reduces and life expectancy increases.
- Ageing populations start to look top-heavy.

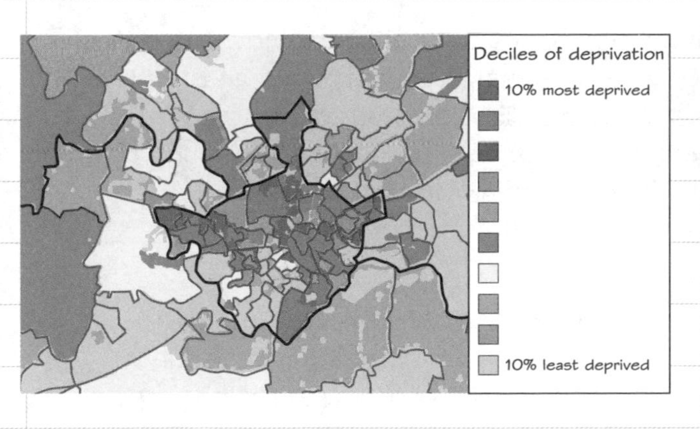

Choropleth maps

Choropleth maps are shaded so that each type of shade represents a particular range of values. The areas used for choropleth maps are usually ones that are used in lots of other ways, like government administrative areas.

Choropleth maps are very good for showing how something varies over a geographical area. One problem with them is that they can suggest abrupt changes between areas where actually changes are much more gradual.

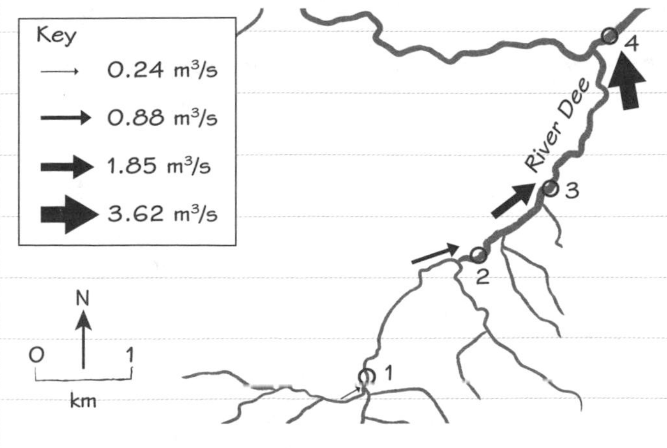

Flow-line maps

Flow-line maps are drawn so that arrows show the direction of flows, and the thickness of the arrows is proportional to the size of the flow.

Flow-line maps are easy to understand and give a clear indication of movement. The relative sizes of the flows can also be clearly seen. However, if lots of flows are going in the same direction the map can start to look very complicated.

Double-headed arrows can be used to show flows in two directions.

Now try this

Study the flow-line map on this page, which was used to present data from a geographical investigation on the River Dee. Suggest what was being measured at the four sites shown. **(1 mark)**

Graphical skills 3

You should know how to interpret dispersion graphs, isoline maps, dot maps, desire lines and proportional symbols (proportional maps).

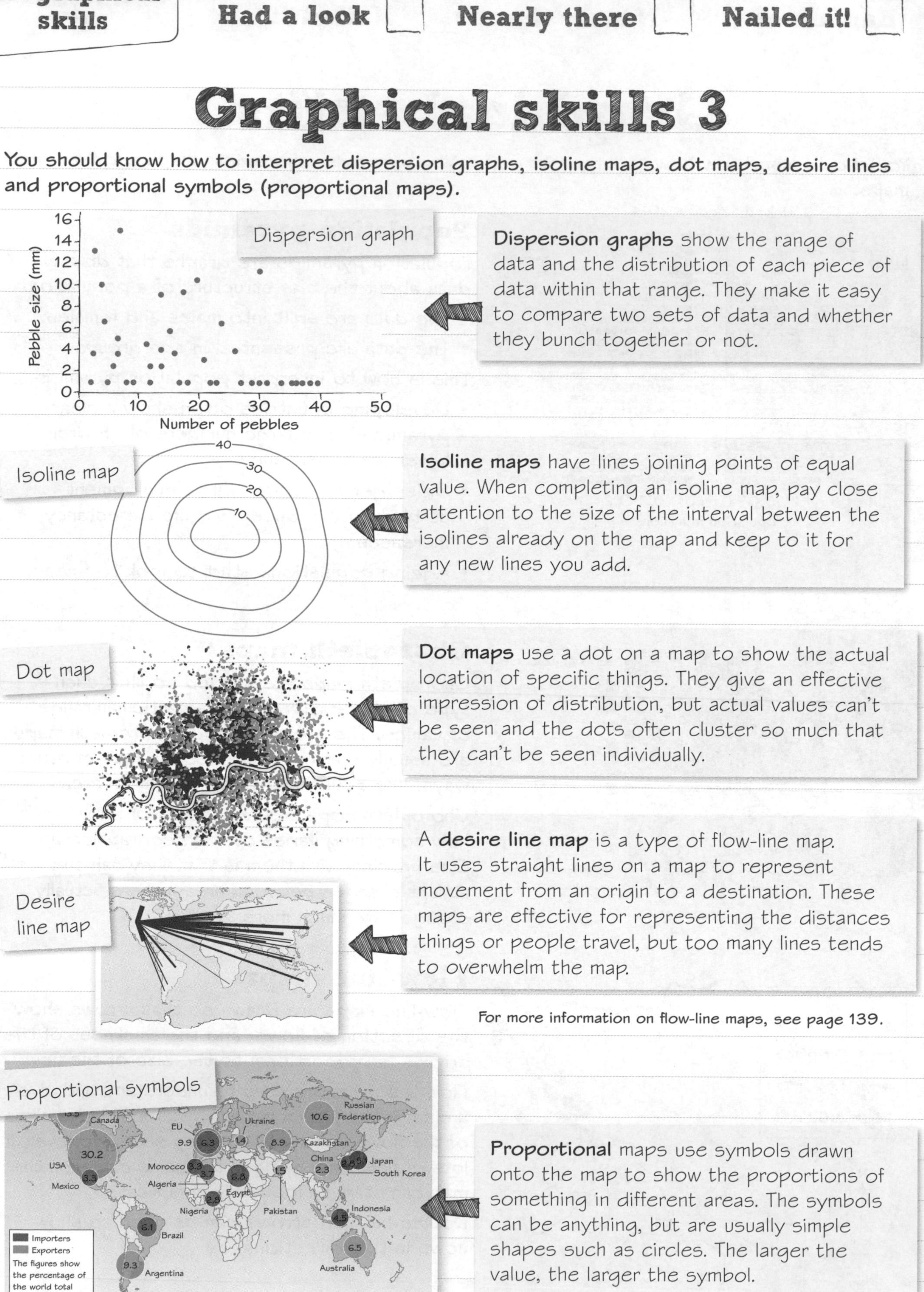

Dispersion graphs show the range of data and the distribution of each piece of data within that range. They make it easy to compare two sets of data and whether they bunch together or not.

Isoline maps have lines joining points of equal value. When completing an isoline map, pay close attention to the size of the interval between the isolines already on the map and keep to it for any new lines you add.

Dot maps use a dot on a map to show the actual location of specific things. They give an effective impression of distribution, but actual values can't be seen and the dots often cluster so much that they can't be seen individually.

A **desire line map** is a type of flow-line map. It uses straight lines on a map to represent movement from an origin to a destination. These maps are effective for representing the distances things or people travel, but too many lines tends to overwhelm the map.

For more information on flow-line maps, see page 139.

Proportional maps use symbols drawn onto the map to show the proportions of something in different areas. The symbols can be anything, but are usually simple shapes such as circles. The larger the value, the larger the symbol.

Now try this

What type of mapping is involved in representing relief by using contours? **(1 mark)**

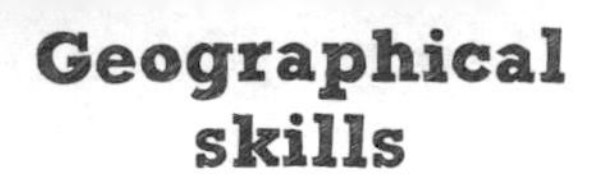

Numbers and statistics 1

You will use your mathematics and statistics skills in specific ways for particular topics in your fieldwork and you may also need them for any of the exam papers.

Proportion and ratio

Proportion – when two values are in direct proportion, as one value increases so does the other, by the same percentage. If one decreases by the same percentage as the other increases, then this is called inverse proportion.

Ratio – indicates the relationship between two quantities, usually in terms of how many times one goes into another.

Equivalent ratios

You can find equivalent ratios by multiplying or dividing by the same number.

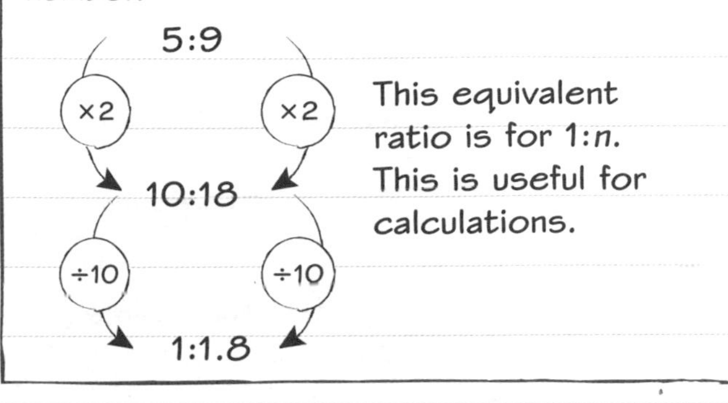

Percentage of an amount

To find the percentage of an amount:

1. Divide the percentage by 100.
2. Multiply by the amount.

For example, 84% of the UK's population of 64 million people live in England.

84 ÷ 100 = 0.84

0.84 × 64 = 53.76

So 53.76 million (53 760 000) people live in England.

One quantity as a percentage of another

To write one quantity as a percentage of another:

Divide the first quantity by the second quantity.

Multiply your answer by 100.

For example, the total energy production in a country is 32 million tonnes of oil equivalent. 8 million tonnes come from coal. To express 8 million as a percentage of 32 million:

8 ÷ 32 = 0.25

0.25 × 100 = 25

So 25% of the country's energy production comes from coal.

Finding percentage increase and decrease

To work out the percentage increase:

Work out the difference between the two numbers you are comparing:

new number – original number = increase

Then divide the increase by the original number and multiply the answer by 100.

To work out the percentage decrease is just the same, except this time the decrease is:

original number – new number = decrease.

Divide the decrease by the original number and multiply by 100.

Now try this

The population of Mumbai's metropolitan area has increased from 8 million in 1971 to 21 million in 2014. What is the percentage increase of Mumbai's population? **(1 mark)**

Numbers and statistics 2

Measures of central tendency help us to manage sets of data by giving us a way of describing them and comparing them easily. You need to know about the three main measures of central tendency – the median, mean and mode – as well as range, quartiles, interquartile range and modal class.

The **mode**: this is the value that occurs **most often**.

4 5 9 7 4 4
The mode of these six numbers is 4

The **mean**: to find the mean you add together all the numbers and then divide by how many numbers there are. Don't round your answer.

4 5 9 7 4 4
The mean of these numbers is 5.5
$4 + 5 + 9 + 7 + 4 + 4 = 33$
$33 \div 6 = 5.5$

The **median**: the median is the **middle value**. First write the values in order from smallest to largest. If there are two middle values, the median is halfway between them.

4 4 4 5 7 9
The median is 4.5

The **range**: the range is the largest value minus the smallest value.

4 5 9 7 4 4
The range of these numbers is $9 - 4 = 5$

Modal class

When data are grouped, the modal class is the group that has the highest frequency.

Study the table opposite to see how this works.

Modal class = 21–25 mm

Size of stones (mm)	Number of stones
1–5	5
6–10	11
11–15	7
16–20	5
21–25	13
26–30	9
31–35	2

Quartiles and interquartile range

The **median** is the middle value: the halfway split in the data. Quartiles divide each half of the data into half: giving us quarters.

The **lower quartile** is the value that divides the lower half of the data into two halves. The **upper quartile** divides the upper half of the data.

The **interquartile range** is the difference between the upper quartile and the lower quartile.

The lower quartile for the data in the table is $(7 + 1) \div 4$ = the 2nd value in the list = 5.

The upper quartile for the data in the table is $3(7 + 1) \div 4$ = the 6th value in the list = 11.

Now try this

What is the interquartile range for the numbers: 2 5 5 7 9 11 13 ? **(1 mark)**

ANSWERS

Where an example answer is given, this is not necessarily the only correct response. In most cases there is a range of responses that can gain full marks.

1. Natural hazards

C A storm surge along a heavily populated coastline

2. Plate tectonics theory

1 Answers could:
- focus on the similarity between the distribution of earthquakes and the location of the margins of the Earth's tectonic plates
- note that earthquakes are most frequent along the major destructive plate margins, such as where the Nazca Plate meets the South American Plate
- point out that the map shows most areas away from plate margins as not having earthquakes.

2 Hotspots are thought to form where a 'plume' of magma rises up from the mantle into the crust. Magma comes up through the crust to create volcanoes. Hotspots provide evidence for plate tectonics theory because, instead of creating just one volcano, hotspots are sometimes associated with chains of volcanic islands, created as the oceanic crust moves over the hotspot. If plates did not move, hotspots would not create chains of volcanic islands, each one geologically younger than its neighbour.

3. Plate margin processes

(i) At destructive margins, one plate is forced under another and pushed down into the mantle, where it melts. This process can lead to magma rising up through the crust to form composite volcanoes.

(ii) At constructive plate margins, plates are moving apart. Magma rises from the mantle to fill the gap between the plates, leading to volcanoes. These shield volcanoes erupt less violently than composite volcanoes.

4. Tectonic hazards: effects

B Secondary effects – an indirect result of the volcano

5. Tectonic hazards: responses

Immediate responses could include one from: the 12-mile exclusion zone being set up around the Fukushima Nuclear Power Plant; 100 000 volunteers being mobilised to help with disaster relief; the Japanese army clearing rubble from roads within two days; 450 000 people being moved to emergency shelters; the repair of electricity, water supply and telephone lines.

Long-term responses could include one from: special development zones set up to attract investment for redevelopment; the US$200 billion rebuilding fund.

6. Living with tectonic hazards

Your reasons could include two from:
- There is fertile farming land around the volcano.
- It could be a long time since the last serious eruption, so people feel it is unlikely that an eruption will affect them.
- People could trust in monitoring equipment on Vesuvius to give the government enough warning of an eruption to evacuate regions at risk.
- People could trust that government preparations for an eruption are good, so that evacuations will be efficient, or that the government will be able to use different methods to reduce the impact of an eruption, such as channelling lava flows away from settled areas.
- People could have strong connections to the area – perhaps their family has lived there for generations, or they have friends and family in the area.
- People might have religious faith and believe they will be protected from a volcanic eruption, or believe that if God has decided they are to die in an eruption, there is nothing they can do about it.

7. Tectonic hazards: reducing risks

One from:
- Government drills in which citizens and emergency services practise what to do when an earthquake occurs means that people know how to reduce the risk of being injured by falling debris, how they should have an emergency kit of food, water, clothing, important documents and medical supplies ready – this could reduce the effects of people having no access to food or clean water, for example, after an earthquake.
- Emergency services will be trained in how to rescue people from collapsed buildings, how to locate settlements that have been cut off and get them supplies, where shelters have been built and how to get people to them. This might not reduce the number of deaths by much, but would reduce the suffering of people left homeless.
- Planned evacuation routes will mean people know where to go and how to get there if they need to leave, reducing congestion and chaos.
- Governments can plan land use and development so that areas at particular risk can be protected with reinforced buildings, or certain types of land use can be prohibited there. This would give important buildings such as hospitals or schools protection to survive a big earthquake.

8. Global atmospheric circulation

B To the west

9. Tropical storms: distribution

They track westwards because of the trade winds which blow to the west. If they reach the belt of the westerlies in the Northern Hemisphere, then these surface winds blow towards the east and this affects the track of the tropical storms, blowing them eastwards too.

10. Tropical storms: causes and structure

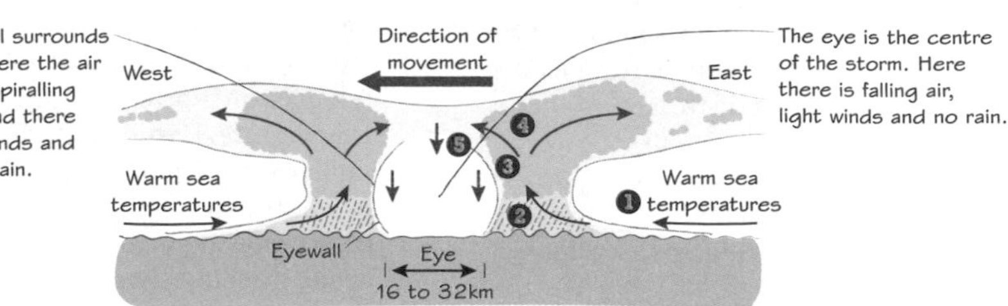

12. Tropical storms: effects

Your two primary effects could include:

- destruction of homes by the storm surge
- destruction of boats, swept onto land by the surge
- roads blocked by debris
- possibly the loss of electricity due to damage to electric cables
- possibly contamination of fresh water supplies by salt water.

13. Tropical storms: responses

It is a long-term response because long-term responses include rebuilding homes that were damaged or destroyed by the tropical storm. It is also a long-term response because long-term responses include attempts to protect an area against tropical storms in the future, or to reduce the impact of future tropical storms. This new house is raised on strong stilts, meaning that its contents will be much better protected from future flooding.

14. Tropical storms: reducing risks

Problems could include two of the following:

- People not knowing in advance where to go to reach an emergency shelter. This could cause traffic jams as people follow each other, instead of their area's evacuation route.
- If people do not have essentials ready, it could delay them leaving their homes, which could put them at risk. They might then need to be rescued by emergency services who would not then be available for other rescues.
- If people do not have essentials ready, they could leave without them. People without medicines could get sick if no other medical supplies were available. If people leave important documents and these get flooded, it would be harder to return to normal life – for example, if they lost house insurance details or their passport.
- Pets are often a problem in evacuation as they get scared and run away. This delays people from leaving and causes them distress.
- People are often reluctant to leave their homes so leave as late as possible. The bad weather then makes travel much more difficult and dangerous.

15. UK: weather hazards

Other examples of extreme weather events in the UK include thunderstorms with lightning strikes and heavy rainfall and heatwaves in which prolonged high temperatures are dangerous for elderly people and people with health problems.

16. UK: extreme weather

Your answer could be along the following lines. The Environment Agency has a strategy of encouraging people in flood risk areas to improve the resilience of their homes against floods. It does this by recommending equipment like door barriers, air-brick protectors and flood kits, which people can buy for their homes.

17. UK: more extreme weather?

A Very useful, as it shows that rain is significantly above average in many regions

18. Climate change: evidence

Possible answers could include one of the following:

- Ice cores are one way in which measurements of temperatures from hundreds of thousands of years ago are possible. The ice sheets in Greenland and Antarctica are thousands of metres deep in places, made up of the accumulation of snow year after year. As snow fell 400 000 years ago, it trapped air particles. Air particles record the air temperatures in a complex way (to do with isotopes of oxygen) and scientists can analyse the air and from it, read the temperature data and other useful things, such as levels of atmospheric CO_2. A higher level of atmospheric CO_2 means an increased greenhouse effect, which means warmer global temperatures, while a lower atmospheric CO_2 can indicate periods of cooler global temperatures.
- A similar process happens with ocean sediments. Some of these are very deep and record layers of sediment laid down year after year for thousands of years. The shells of tiny marine creatures are made up of calcium carbonate which, like the ice cores, record the variations in oxygen isotopes that tell us what the temperature of the sea was when they were formed.
- Fossil records can also help indicate temperatures. For example, fossil tree rings record the length of the growing season each year, which is longer or shorter according to temperature. Fossil corals do a similar sort of job. It is also possible to tell from fossil pollen whether the plants growing in a region are warm-loving or cold-tolerant plants.

19. Climate change: possible causes

Any two from: carbon dioxide, methane, nitrous oxide

20. Climate change: effects

Here is one possible approach to answering this question: there are plenty of others.

- Consider some positive effects of climate change: for example, people may not need to spend so much money on heating during winter months and farmers in northern countries are likely to be able to grow crops for longer in the year, and introduce new crops, which could increase the amount of food available and increase profits made by these farmers.
- Then consider negative effects and indicate that these will affect far more people around the world than will benefit from climate change. Perhaps also say that the people who will suffer most would be people who are already living challenging lives in poorer areas of the world – for example, people who are at risk of increased droughts, putting greater pressure on available water resources and leading to further desertification of semi-arid areas, as in the Sahel. People who live in coastal regions – around 23 per cent of the world's population – will be affected by sea level rises, and in many parts of the world there will be increased risks of coastal and river flooding. Increased temperatures will spread diseases into new regions. The impact on farming is likely (globally) to be more negative than positive.
- As a final point to support the statement, even those areas and people who benefit from climate change in some ways will probably end up dealing with more negative consequences than positive ones. Even if people do not live in areas at risk of flooding or in low-lying coastal areas, they will need to pay extra taxes to pay for the costs of protecting those who do. Even if people are not affected directly by desertification, water shortages and crop failures, it is likely that they will be affected by the impacts of increased movements of people away from badly-impacted countries. And everyone is likely to be affected by the greater risk of extreme weather events.

6-mark questions have got three mark bands. The first band is **basic**, worth 1–2 marks, where only one or two points are made, there is not much or any supporting evidence or detail, and the answer is not easy to understand. **Clear** answers are worth 3–4 marks: the points are made clearly, the points made are relevant and some points are supported. At the top level, **detailed** answers are worth 5–6 marks. These answers are clear and make relevant points that are supported by detail.

21. Mitigating climate change

Mitigating climate change means reducing the causes of climate change.

22. Adapting to climate change

B Planting trees to increase carbon storage; **C** Installing solar power on government buildings

23. A small-scale UK ecosystem

Nutrient cycling is the process by which nutrients are released from dead organisms by decomposers and returned to the soil, and then taken up again by producers, eaten by consumers, who then die – and the process repeats itself in a circle.

24. Ecosystems and change

Nutrients in tropical rainforest soils are usually taken up rapidly by the abundant, fast-growing rainforest vegetation. Nutrients are constantly returned to the soil by falling leaves and when trees and other organisms die. When trees are removed from the ecosystem, this balance is upset. Nutrients are not returned to the soil and the nutrients that are left in the soil are rapidly leached away by the frequent and heavy rainfall.

25. Global ecosystems

1 C Temperate ecosystem
2 The tundra large-scale ecosystem is located at very high latitudes, within the Arctic Circle. It is not found in the Southern Hemisphere because of the distribution of the Earth's land masses.

26. Rainforest characteristics

Rainforest trees grow very tall (30–45 m). One other physical characteristic of these trees is that they have (possible answers) shallow roots/broad leaves/leaves with drip tips/thin trunks/ buttress roots.

27. Interdependences and adaptations

One of the physical conditions of the tropical rainforest is that the nutrients are found in the top layer of the soil, as they are leached out deeper into the soil. Consequently, trees have shallow roots. Because they grow so high, rainforest trees have evolved buttress roots to increase their stability.

28. Deforestation: causes

Your answer will depend on your case study. The main causes of rainforest deforestation are commercial farming (including clearing the forest to grow biofuels such as sugarcane), subsistence farming (when poor farming families clear enough land to grow the crops they then live on), population growth (linked to more people clearing land for subsistence farming), logging (including selective logging, which still promotes deforestation), and mining and energy development (for example, flooding forests to create HEP reservoirs).

29. Deforestation: impacts

Your answer will depend on your case study. The main impacts of rainforest deforestation are soil erosion and the contribution to climate change (trees take in carbon dioxide from the air and store it); don't forget that impacts can be positive too – for example, economic development: deforestation creates land for farming, land for energy production, land for building settlements on or roads and railways to link settlements.

30. Rainforest value

B Carbon sequestration

31. Sustainable management

Tropical rainforests tend to be located in poorer countries and logging, commercial farming and mineral extraction from cleared forest areas all represent good ways for poorer countries to earn money that they could then use for development. Because many poorer countries have large international debts, strategies have been suggested and implemented by which poorer countries agree to conserve areas of their tropical rainforests in a return for a reduction in their debt payments. These deals are known as conservation swaps.

32. Hot deserts: characteristics

The subtropical high-pressure zones occur because dry air is descending. Clouds do not form under these high pressure conditions, giving hot deserts clear skies in the day and at night. Although the Earth's surface stores a lot of heat from the daytime high temperatures, at night this heat is quickly lost because there are no clouds to trap the radiated heat in the atmosphere. Consequently deserts are very cold at night – as low as 0°C.

33. The hot desert ecosystem

Your answer is likely to make one of the following points.

- Because hot deserts are low-nutrient ecosystems, plants grow very slowly here. This means that any damage to plants takes a long time to be repaired.
- The species that survive in hot deserts are highly specialised and adapted to their environment. This means that, if the environment is altered, they are likely to struggle to survive.
- Because biodiversity is low in hot deserts, there are no other species ready to take over if one species is affected, for example, by disease.
- Because biodiversity is low in hot deserts, species are often highly interdependent. If one species becomes scarce for some reason, other species will be badly affected, too.

34. Development opportunities

This answer will depend on the case study you have studied. You will have studied both opportunities and challenges in your case study. Note that although this question is only asking about opportunities, it does say 'to what extent', meaning that you might be able to use some of your knowledge of challenges to explain the difficulties in development reaching its full potential.

For example, if your case study includes the opportunity for development of solar power it would be good to describe the extent of development so far (the amount of power generated would be a great statistic to use) and explain why the physical conditions of the desert mean that there is lots of potential for solar power development. But to cover the 'to what extent' issue, you could consider factors that might limit this opportunity: for example, the dust and dirt of the desert environment that covers the solar panels and reduces their efficiency, or the damage done to solar panels by strong desert winds, or the long distance between where solar power is generated and where there is demand for that power.

The same idea of limiting factors on opportunities for development can be used for farming (where is the water coming from) and tourism (where is the water coming from, how far do tourists have to travel to reach the location).

9-mark questions like this one use levels to work out the marks for the answer as a whole. There are three levels: **basic** answers, which have a maximum of 3 marks, **clear** answers (marks of 4–6) and **detailed** answers (7–9 marks). 9-mark questions need answers that combine your knowledge (detailed information), understanding (geographical concepts) and evaluation (weighing up the 'to what extent' part of the question).

35. Development challenges

This answer will depend on the case study you have studied. Notice that the question is looking for evidence that water supply is a challenge, so here you can use examples about the challenges of water supply from your case study: perhaps to do with the amount of water that is needed for tourist developments in your case study hot desert, or for agriculture and why this is a challenge given available supplies. You will need to make sure you use your case study evidence to show why water supply is a challenge: don't just write down water supply information.

Although the question asks specifically for evidence about water supply challenges, notice that it says 'the biggest challenge'. That means you can also write about other challenges for developing hot desert environments, but only in ways that show that water supply remains the biggest challenge. For example, you might be able to use evidence from your case study that inaccessibility is a major problem for development in your case study hot desert, but that the main challenge of this inaccessibility is getting enough water to the inaccessible location. Or you might talk about extreme heat being a problem for solar power generation in your case study location, but again show that water supply is critical for this as the solar power cells have to be cooled using water.

6-mark questions have got three mark bands. The first band is **basic**, worth 1–2 marks, where only one or two points are made, there is not much or any supporting evidence or detail, and the answer is not easy to understand. **Clear** answers are worth 3–4 marks: the points are made clearly, the points made are relevant and some points are supported. At the top level, **detailed** answers are worth 5–6 marks. These answers are clear and make relevant points that are supported by detail.

36. Desertification: causes

Your answer could include one of the following:

- More people means more farming of dryland soils, which can lead to soils becoming exhausted and degraded.
- More people could lead to the development of irrigation systems, which could have the impact of increased productivity in farming.
- More people could mean more animals grazed on dryland areas, which can lead to vegetation being eaten and the soil becoming loosened, dried out and vulnerable to erosion and desertification.
- More people could lead to stress on the available water supply in the area, which might mean rivers and streams become more seasonal, putting stress on animals that depend on these sources.
- More people might mean an increase in demand for fuelwood, both for locals and to be converted into charcoal and sold in cities for fuel. This has the impact of degrading soils by clearing trees and shrubs that help bind the soil.

37. Desertification: reducing the risk

Your reason could include one of the following:

- Trees would provide shade that would protect the crops from extreme heat; keeping the crops alive helps maintain soil integrity.
- Tree shade could reduce evaporation of water from the surface of the soil, which could reduce salinisation.
- Trees provide some shelter from strong winds, which could help protect the soil from wind erosion.
- Tree roots would help bind the soil, making it less prone to erosion from wind and water.
- Trees would help add nutrients and structure to the soil from their leaves when they fall.

38. Cold environments: characteristics

Around 225 mm

39. The cold environment ecosystem

Your answer is likely to make one of the following points.

- Because cold environments are low-nutrient ecosystems, plants grow very slowly here. This means that any damage to plants takes a long time to be repaired.
- The species that survive in cold environments are highly specialised and adapted to their environment. This means that, if the environment is altered, they are likely to struggle to survive.
- Because biodiversity is low in cold environments, there are no other species ready to take over if one species is affected, for example, by disease.
- Because biodiversity is low in cold environments, species are often highly interdependent. If one species becomes scarce for some reason, other species will be badly affected too.

40. Development opportunities

- This answer will depend on the case study you have studied. You will have studied both opportunities and challenges in your case study. Note that although this question is only asking about opportunities, it does say 'to what extent', meaning that you might be able to use some of your knowledge of challenges to explain the difficulties in economic development reaching its full potential.
- For example, if your case study includes the opportunity for development of energy resources (oil and gas extraction) in a cold environment it would be good to describe the extent of development so far (the amount of oil or gas extracted would be a great statistic to use) and explain why changing environmental conditions might be making this development easier and cheaper than before. But to cover the 'to what extent' issue, you could consider factors that might limit this opportunity: for example, the environmental concerns about increasing energy development in cold environments.
- The same idea of limiting factors on opportunities for economic development can be used for tourism (how far do tourists have to travel to reach the location).

41. Development challenges

This answer will depend on the case study you have studied. Notice that the question is looking for evidence that extreme temperature is a challenge, so here you can use examples about the challenges of very low temperatures from your case study: perhaps to do with the physical risk of low temperatures to human health, the impact low temperature has for heavy snow, for creating uncertain icy terrain and the problems extreme temperatures cause for equipment. You will need to make sure you use your case study evidence to show why extreme temperature is a challenge for those trying to develop the cold environment: do not just write down extreme temperature information.

Although the question asks specifically for evidence about extreme temperature challenges, notice that it says 'the biggest challenge'. That means you can also write about other challenges for developing cold environments, but only in ways that show that extreme cold remains the biggest challenge. For example, you might be able to use evidence from your case study that inaccessibility is a major problem for development in your case study cold environment, but that the main challenge of this inaccessibility is getting enough energy to the inaccessible location to provide sufficient heating to keep the development safe from extreme temperature hazards. Or you might talk about provision of building and infrastructure creating difficulties for developing cold environments, but again link this to the need for buildings to provide adequate protection from extreme temperature, or for infrastructure to be able to survive the permafrost caused by very low temperatures.

42. Fragile wilderness

Your two ways could include the following:

- As a place where humans can experience a natural environment, rather than one that has been changed by other humans
- For its biosphere services: for example, reflection of radiation from the Sun during winter when the area is snow-covered, which helps to reduce global temperatures
- As a natural habitat: tundra environments like this host hundreds of species of migrating birds during the summer months, when the lakes and marshes of this environment produce huge numbers of insects

- As a natural ecosystem that can be used to monitor the effects of climate change
- As an ecosystem that can be studied in its natural state, with minimal human impacts
- Possibly as an environment in which First Nation peoples live: the original inhabitants of Canada, who developed sustainable ways of living with this environment

43. Managing cold environments

Your reason is likely to make one of the following points.

- Too many visitors could frighten local wildlife (the time when tourists are able to visit Antarctica is also the breeding season for penguins and other Antarctic bird and mammal species).
- The Antarctic environment is fragile. Lots of tourists could damage the environment by erosion (there are only a few good spots to land in the most accessible part of Antarctica, so tourists are concentrated in just a few places).
- The more tourists there are, the higher the risk of visitors contaminating local habitats (for example, a tourist may have plant seeds from their own country on their boots or clothing, which might introduce non-native species into the polar ecosystem).

44. Physical landscapes in the UK

Your two differences might include the following:

- While the upland landscape has rugged relief, the lowland landscape has gentle relief.
- While the upland landscape has bare rock faces, the lowland landscape has grass-covered hillsides.
- While the upland has very steep valley sides, the lowland landscape has less steep valley sides.
- While the upland has a mass of mountainous terrain, the lowland landscape has hills and then a very wide, flat valley floor.
- There is a mountain lake (a tarn) in the upland landscape, which feeds a steep-sided mountain stream down the mountainside to the valley floor. In the lowland, there is no surface water visible in the photo, but the map extract shows streams running across the flat valley floor.

45. Types of wave

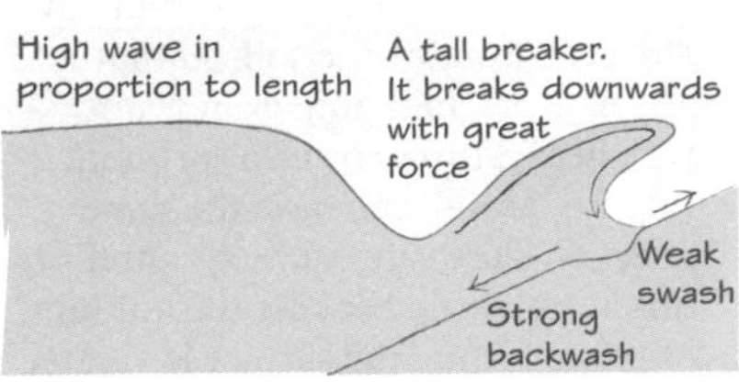

Destructive waves have a high frequency (10–14 waves per minute) and short wavelength, producing a plunging, high-energy wave with a strong backwash, which causes erosion.

46. Weathering and mass movement

The chalk could be affected by chemical weathering – like limestone, chalk is dissolved by carbonation. Biological weathering is not on the specification, but you would still be credited for giving it as a type of weathering occurring at the top of the cliff. It is not likely that exfoliation is involved as temperature ranges in Devon are not regularly high enough.

47. Erosion, transport, deposition

Prevailing wind

48. Erosion landforms

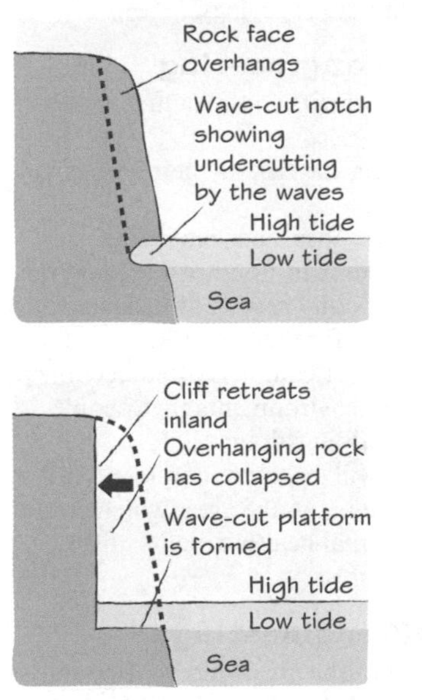

The sea erodes the base of the cliff, forming a notch. Over time, the notch gets bigger and the cliff is undercut, retreating inland and eventually collapsing. Debris from the collapsed cliff is washed away by the sea, exposing a wave-cut platform underneath.

49. Deposition landforms

Obstacles cause the wind to slow down, reducing its energy. Like water and deposition, when wind has less energy, it will deposit the material it no longer has enough energy to transport. As the sand dune gets bigger, it forms more of an obstacle and leads to more sand being deposited. (Very strong winds can blow the sand away from the obstacle, however, and this is why sand dunes that are not fixed by vegetation can be temporary landforms.)

50. Coastal landforms

Groynes (named on the map extract at grid reference 183908 and visible on the aerial photo from the distinctive triangular shapes they produce as sand builds up behind them)

51. Hard engineering

Abrasion (also known as corrosion)

52. Soft engineering and managed retreat

Your reason is likely to make one of the following points.

- Although it is cheap to make holes in existing coastal protection to let the sea through, and further maintenance is not often required, there are other costs involved in coastal realignment, including loss of earnings for farmers and businesses on the land affected (for example, static caravan parks) and compensation for the loss of land.
- Coastal realignment is also often unpopular with local communities, who generally prefer that the coast was protected.

53. Coastal management

This will depend on the example you have studied, but could include the following points.

Reasons for coastal protection (usually people want hard engineering) include: to prevent people from losing their homes; to prevent people from losing their businesses; to prevent loss of wildlife habitats; to prevent loss of transport links (for example, road and rail links); to prevent loss of historic sites (for example, prehistoric archaeology)

Reasons against coastal protection: the expense of building sea defences; the continuing expense of maintaining defences/repeating beach nourishment or beach reprofiling every year or two; building sea defences in one location can make problems worse elsewhere; building sea defences can damage the environment

54. River valleys

The cross profile of the river shows you a cross section of the river channel and valley, showing the changes in the shape of the river channel, width and depth. The long profile shows how the gradient of the river changes as you move downstream.

55. River processes

Your reason is likely to make one of the following points.

The velocity (speed) of the water is reduced (the bigger the sediment particle size, the higher the velocity needed to transport it):

- when the river enters a lake, reservoir or the sea, or at the base of a waterfall
- when the river flows against an obstacle
- when velocity is reduced by friction on the river banks or bed
- when the river's discharge (the amount of water in the river) is reduced: for example, during a drought.

The amount of sediment carried by the river (sediment load) is increased above the capacity of the river to carry it.

56. Erosion landforms

Your characteristics would include two of the following: river, plunge pool, waterfall, cap rock, overhang, steep-sided gorge

57. Erosion and deposition landforms

567202

58. Deposition landforms

Your answer should identify the features as levées and your explanation should focus on how they are formed rather than describing what they look like. Your explanation is likely to include the point that when a river floods, it spreads out and loses energy. The heaviest/coarsest material that it is carrying is deposited first, on the sides of the river. Over long periods of time, following many repeated floods, this deposition of the heaviest particles on the banks of the river creates levées – natural embankments along both sides of the river.

59. River landforms

Your symbol meanings should include two of the following:

- House symbol = building of historic interest
- Oak leaf = National Trust
- Cross in a broken circle = Cadw – Welsh National Heritage
- Boat on a slope = slipway
- i sign = information centre
- PC = public convenience (toilet)
- Footprint = walks/trails
- Square shape in a circle = World Heritage site

The little blue 'w' represents a well or a spring; it is not a tourist or leisure symbol.

60. Flood risk factors

The points you make will include some of the following. You need to make **one** point about precipitation and explain how it increases the risk of flooding and **one** point and an explanation for geology (rock type).

- Precipitation increases flood risk when heavy, continuous rain saturates the soil. When the soil is saturated, more water cannot infiltrate the soil and surface runoff increases. Surface runoff means water reaches the river more quickly, rapidly increasing discharge and increasing the risk of floods. The same effect can also occur when long periods of rain have already saturated the soil, so that any new precipitation, even if not very heavy, turns to surface runoff. Intense rainfall can have the same effect of increasing surface runoff on unsaturated soil because the amount of water hitting the surface is greater than the rate at which the soil can absorb it. If the soil is bare, and also if it has been compacted, then this effect is increased. Snowmelt also leads to an increase in surface runoff and this will increase the flood risk, especially if snow melt is rapid.
- Geology (rock type) increases flood risk when the underlying rock of a river's catchment is impermeable and close to the surface, as this reduces infiltration and means surface runoff is increased. Rocks like granite and shale are impermeable and these are common rock types in the UK's mountain areas, where soils are also often thin. Clay soils are also impermeable, which increases flood risk in many lowland valleys. However, clay soils support a lot of vegetation, which reduces surface runoff. Permeable rock types like limestone permit infiltration and this significantly reduces flood risk in areas with limestone geology.

61. Hard engineering

Your two benefits could include the following:

- It reduces the risk of their homes flooding.
- It reduces insurance payments (people in high-flood risk areas often have difficulty getting affordable insurance for their homes).
- Flood relief channels often become pleasant environments that people enjoy walking along.
- People will feel more confident and happy living in the area, which could have mental health benefits and reduce stress.

62. Soft engineering

The main problems would be if people were not aware of the website or ways to receive flood warnings on social media, or if people did not have a computer or a phone that could receive social media alerts – particularly elderly people. Also, during floods, power can cut out and therefore people cannot use computers (this happened in Lancaster in 2015). Other problems could include: if people are not skilled at reading maps or not clear on what the warning means for them; if people are confused about what the different flood warning levels mean in practice and what they should do as a result.

63. Flood management: example

This answer will depend on the example you have studied in class.
Here are some example answers for the Pickering (Yorkshire) flood management scheme. You only need to make one point for each issue in your answer.
Social issue – in Pickering, it was local residents themselves who wanted to use environmentally friendly soft-engineering strategies to reduce flood risk, rather than put pressure on local government for a hard-engineering strategy. As a result, the local community is very supportive of the scheme and are proud of how well it has worked.
Economic issue – flooding in Pickering in 2007 caused £7 million of damage, and 2007 was the fourth time the town had flooded since 1999. The Environment Agency said that the best solution was a £20 million concrete wall in the town, but local residents thought this would damage tourism, which is a main source of the town's income. Then the Environment Agency said it could not justify spending £20 million in Pickering anyway. The soft-engineering strategies cost £2 million in total, have proved to work well and do not impact negatively on tourism.
Environmental issue – the soft-engineering strategies in Pickering were inspired by natural processes that reduce flood risk. They included tree planting, which provided additional woodland habitat for local wildlife and 'leaky' dams, deliberately clogged with debris, which let normal river flows through but hold back high discharge levels. The leaky dams mimic natural river conditions. An earth bund or embankment was also built (which can retain 120 000 m^3 of floodwater), so that the river's natural flood plain could store floodwater rather than it going into the town.

64. Glacial processes

Smoothing of the bedrock happens when the ice next to the bedrock carries small rock particles. These small particles act like sandpaper on the bedrock, grinding away irregularities to produce a smoother surface. When the ice near the bedrock carries sharper, larger rock fragments, these scratch along the bedrock as the ice moves, causing striations.

65. Erosion landforms 1

A tarn or lake

66. Erosion landforms 2

Your diagram should look something like this.

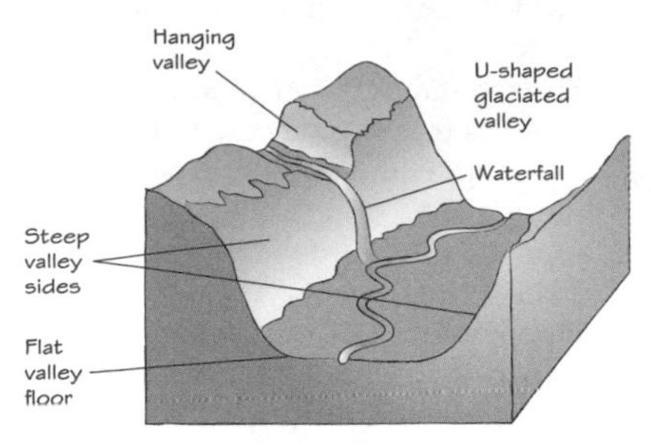

67. Transportation and deposition landforms

These are end moraines left at the snout of the glacier as it retreats up the valley. There may be several of these: the terminal moraine is the end moraine that marks the farthest extent the glacier reached.

68. Upland glaciated area: example

East across Lake Buttermere

69. Activities and conflicts

Your two ways could include the following:
Tourism can conflict with farming because of tourists leaving gates open or trampling crops, or when tourist dogs frighten or kill livestock (for example, lambs, sheep). Farming can conflict with tourism when farmers make it difficult for tourists to follow rights of way over their land: for example, by failing to repair stiles (places to climb over walls) on their land or not clearing brambles or nettles away from footpaths. Farm vehicles can get in the way of tourist traffic, holding up traffic on winding upland roads.

Tourism can conflict with forestry when plantations block tourist views across a scenic upland area or when an area of the plantation is felled, leaving bare, churned-up ground, tree stumps, broken branches and sawdust piles. Conifer forests are also dark, monotonous places to walk through without much wildlife. Tourists may not enjoy this experience as much as walking through the open upland areas. Forestry can also sometimes be noisy and sometimes involves culling animals, especially deer, which is not compatible with almost all tourist activities. Tourists can also cause problems for forestry if, for example, landowners are refused permission to develop plantations in upland areas for conservation reasons or because it would impact on tourist numbers by changing the landscape.

Tourism conflicts with quarrying because tourists are not attracted to the noise and other environmental and landscape impacts of quarrying. Lorries carrying processed stone are not welcomed by tourists who are walking along upland area roads or tourists who get stuck behind lorries in their cars. Conservation of the upland area may mean that quarrying operations have to follow restrictions to minimise their impacts on tourism, such as restricted times when blasting can occur or when lorries can transport materials from the quarry. Conservation to preserve the attractiveness or special character of the area for tourists may mean that existing quarries are not allowed to expand and new quarrying sites are not permitted in the upland area, which limits the potential of quarrying and reduces the numbers of jobs it can create for local people.

70. Tourism

Your answer will depend on the glaciated upland area you have studied. Here are two example answers using the Lake District.

- In the Lake District, Fix the Fells is an organisation led by the National Trust that restores and repairs upland footpaths that have been damaged by erosion or are at risk of being damaged.
- The Go Lakes Travel initiative has provided £7 million for cheaper, cleaner and better integrated public transport in the Lake District to reduce congestion caused by visitors travelling everywhere in their own cars.

71. Global urban change

11 per cent. [To calculate percentage change, you need to divide the actual increase by the original value and then multiply it by 100. For this question, the actual increase is 3.15 billion minus 2.84 billion = 0.31 billion, and the original value to divide it by is 2.84 billion.]

72. Urbanisation factors

16.3 per cent. [To calculate a percentage increase, first work out the difference between the two values – in this case 8.6 million and 10 million, so 1.4 million. Then divide 1.4 by 8.6 and multiply the result by 100 to get your percentage increase: 16.27. Rounded up to one decimal place, this is 16.3 per cent.]

73. Non-UK city: location and growth

This answer will depend on the city you have done for your case study. Make sure you include information about the population size now, how it has changed over time, the rate of urban growth and causes of the growth, especially the contribution of migration and natural increase. Be sure to include facts and figures from your case study in your answer.

74. Non-UK city: opportunities

This answer will depend on the city you have done for your case study. The sorts of social opportunities that bring migrants to growing cities in LICs and NEEs involve access to health and education services and access to resources like water supply and energy.

75. Non-UK city: challenges 1

This answer will depend on the city you have studied, but make sure to include facts and figures from your case study city. Your answer might include the following:

- Evidence about overcrowding in slums or squatter settlements (population density)
- Information about where squatter settlements have been located: for example, on steep hillsides (as in parts of Rio de Janeiro or Mexico City) or besides lakes and rivers (as in Lagos) or along railway lines (as in Mumbai), and why these locations cause challenges for residents
- Evidence about problems accessing clean water: for example, what percentage of residents of your city have access to clean piped water; how many people there are per toilet in slum or squatter settlements
- Information about causes of water pollution, perhaps from a lack of sanitation, industrial pollution or pollution during river floods or coastal flooding.
- Energy supply information: for example, what people use for cooking with (bottled gas, charcoal) and its impacts, whether there are frequent blackouts due to demand for electricity exceeding supply

76. Non-UK city: challenges 2

This answer will depend on the city you have done for your case study. Your answer might include the following:

- Information about traffic congestion and traffic management plans: for example, city government schemes to integrate transport through bus transit schemes (as in Lagos), or monorail schemes (as in Mumbai) or toll road schemes and metro expansions (as in Rio), or restrictions on car numbers entering the city on specific days (as in Mexico City)
- Air pollution information: for example, evidence on the numbers of deaths linked to air pollution per year, and information about city government controls on polluting industries or penalties for using older, more polluting vehicles
- Waste disposal problems: it is common in LIC and NEE major cities for people to have to live alongside city waste dumps, which release unpleasant and sometimes harmful fumes. City governments have sometimes tackled these problems by moving waste disposal further out of the city and restoring waste dumping areas through ground sealing and re-landscaping

approaches: for example, in Mumbai the Gorai Garbage Site Closure Project won an award for sustainable urban development and property prices increased significantly in the area because what had been a massive waste dump became an attractive park.

- Water pollution issues: for example, evidence on sources of water pollution (from sewage, illegal waste dumping, industrial pollution). Many major cities are located by river estuaries and the rivers coming into the city are often already heavily polluted. City government solutions include building new sewage treatment plants. Charities advise slum and squatter settlement residents on how to increase their rainwater storage, and communities may form pressure groups to convince city governments to extend clean, piped water supplies to their communities.

77. Planning for the urban poor

This answer will depend on the example(s) you have studied. For example, if you have studied urban planning in Curitiba, Brazil, your answer might make some of the following points.

- Public transport is very successful – used by 85 per cent of the population and heavily relied on by poorer people. The system is fast, cheap and efficient. The buses run frequently, some every 90 seconds. The system transports 2.6 million people each day. It has helped improve people's lives by reducing congestion on the roads. It also allows people to live in cheaper, outlying areas but still be able to afford to commute to jobs in other parts of the city. The buses use alternative fuels such as natural gas, which creates less air pollution, improving quality of life for everyone in the city.
- Curitiba has preserved its green areas, with 28 parks and wooded areas making the environment more attractive for people. Builders get tax rebates if their projects include green space and new lakes absorb floodwaters, which were a problem in the past.
- People living in low-income areas bring their rubbish bags to centres where they swap them for bus tickets and food. This both helps with waste disposal problems in slums and squatter settlements and also makes transportation easily affordable for the urban poor. Children can also exchange recycled waste for school supplies, food and chocolate – reducing waste (less litter and disease) as well as helping education.

78. Urban UK

The uneven population distribution of the UK could be described using some of the following points.

- There are two main areas of particular high density – the south-east and from the Midlands to the north-west of England. These areas are dominated by large cities where economic activity is the greatest (London, Birmingham, Manchester).
- Sparsely populated areas are located where relief is a prime factor, such as central Wales (Snowdonia) and the Highlands of Scotland.
- Areas of moderate population density surround the densely populated areas and the regional cities mostly centred on the coast (Cardiff, Newcastle, Edinburgh).

79. UK city: location and migration

This answer will depend on your case study city. Here is an example answer for Birmingham.

Since 2004, Birmingham's population has increased by almost 100 000 people (a 9.9 per cent increase). International migration has been significant in this strong growth rate, both because of more people coming from other countries to live in Birmingham (there has been a 6 per cent increase in the proportion of Birmingham residents who are from India, Pakistan and other Asian countries) and because migrant families have contributed to a high rate of natural increase in Birmingham, because migrants have usually been young adults who have had families since moving to Birmingham.

80. UK city: opportunities

This answer will depend on your case study city. Here are examples of what you might include.

Social opportunities:

- Opportunities resulting from an increasingly youthful population: for example, the development of lesbian, gay, bisexual and transgender (LGBT) quarters in the city.
- Opportunities resulting from an increasingly diverse population: for example, different cuisines to try, new cultural events to attend.
- New sporting developments: for example, following city or national funding for a new stadium or infrastructure for a sporting competition, such as the Olympics in London.
- Changes in retailing as a result of the move towards shopping as an 'experience'.
- Expansion of schools so they can accommodate more children, and expansion of hospitals so they can cope with any increase in the numbers of people needing medical attention.

Economic opportunities:

- How jobs have changed over time, from industrial (secondary) to services (tertiary and quaternary sectors).
- Hi-tech industries in your city, or changes from new media industries or research and development industries.
- Opportunities provided by big new retail developments.
- Opportunities from changes in city transport, especially the development of integrated transport systems.

81. UK city: challenges 1

This answer will depend on your case study city. Here is an example answer for Birmingham.

Economic change in Birmingham has involved the decline of the car industry, which previously employed large numbers of people in the city, causing high unemployment. New industries have developed, including finance, IT and advanced manufacturing – industries demanding high skills from employees and job applicants.

The inner city areas suffered from high unemployment following deindustrialisation, and dereliction and growing social problems in the inner city (and some suburban areas) meant that new industries avoided locating in the inner city areas. There were also issues that the older buildings in these areas were not always suitable for the new industries. People in the inner city had to travel to find jobs, and if they succeeded in finding jobs they usually then left the areas of deprivation as soon as they could.

International migration has also been involved in creating inequalities, as the only place that new arrivals in Birmingham could afford to live was in the inner city. The jobs available in deprived areas tend to be low paid and often temporary. Older people have also sometimes been 'left behind' by the changes in Birmingham: while many young people have been able to develop the skills needed to take up opportunities in tertiary and quaternary jobs, older people, who previously worked in the car industry, have often struggled to adapt. This is especially true for older, lower-skilled people.

The rapid growth of the city has also stretched city budgets so that the city government has not always been able to afford to make improvements to deprived areas of the city which might make these areas more attractive to businesses.

82. UK city: challenges 2

Your answer will depend on your case study city. The prompts about urban sprawl in the spider diagram on page 82 should help with your answer.

83. UK urban regeneration

Your one advantage could be one from:

- more green space for leisure and exercise
- property prices likely to increase
- wider range of shops and services: for example, new leisure centres, libraries, colleges, university campuses
- development may include affordable homes
- local council will get a lot of money from sale of council property and from higher rents in the redeveloped area
- social problems of deprived urban areas are removed or reduced
- older city areas prioritise cars over pedestrians: regenerated areas are more friendly for pedestrians and cyclists with wider paths, cycle lanes, traffic calming measures and green walkways and leisure walks
- new businesses move to regenerated areas, bringing new jobs
- bigger malls with a wider choice of shops.

Your one disadvantage could be one from:

- rental prices increase too much for residents to afford to live there any more
- familiar shops and services (for example, pubs, cafes) close down to be replaced by new and more expensive shops and services
- regularly increasing property values means investors buy houses or flats in the area only to leave them empty until they want to sell them again
- new jobs may require skills and qualifications that existing residents do not have
- regeneration may have taken place without consultation with the community, so existing residents do not feel they had a say in changes to their neighbourhood
- economic problems may mean that new business and new services do not move to the new development in the way that was expected, wasting council money that could have been spent helping local people
- the regeneration of one area may move social problems to another area as poorer residents who have been forced out by higher prices move to an area where they can still afford to live.

84. Sustainable urban living

Your two ways are likely to include: solar panels, insulation, recycling schemes. You might also suggest: terraced housing rather than detached or semi-detached (reduces amount of heat loss through external walls); double-glazed windows and doors; location near public transport (for example, bus stop, tram stop), construction on brownfield site.

85. Urban transport strategies

1 The two ways identified in the information are:
  - reducing the numbers of vehicles travelling inside the congestion charge zone during busy times
  - using money raised from the congestion charges to invest more in public transport and in cycling schemes.

2 This answer will depend on your case study. For a **discuss** question you would usually look at advantages and disadvantages of the urban transport strategy or strategies. So, for example, your advantages might include: statistics for reducing the number of vehicles in the congestion-controlled part of the city, statistics on reducing air pollution in the city, increased investment in public transport and other forms of transport that reduce congestion, such as cycling, walking. Disadvantages could include: commuter transport (for example, trains) becoming very crowded and unpleasant for commuters; increased costs for businesses if the scheme makes it more expensive or slower to bring vehicles into and out of the city; possibility of businesses moving out of the city to a new location that is cheaper; some parts of the city having better public transport than others – producing inequalities.

86. Measuring development 1

You would expect a positive correlation between HDI and GNI per head because, as income increases per head, so too should the things that HDI measures, such as life expectancy, literacy and education.

87. Measuring development 2

The human development index measures indicators like life expectancy and educational opportunities as well as GNI per head. The governments of some LICs have decided to make education a major focus, which can explain why some LICs have high human development but relatively low economic development.

88. The Demographic Transition Model

Stage 4

89. Uneven development: causes

You should aim to make two developed points for this answer. Examples include:

- prices – prices for primary products go up and down very rapidly as they are traded in HICs and NEEs, and often they can drop sharply just as a country has spent its resources building up production
- competition – there is often strong competition between LICs producing the same primary product which drives prices down
- processing – the most profit is made in processing raw materials into manufactured products because they are value-added. The HICs and NEEs that do this are very keen to stop any other countries from joining in; the processing countries also work hard to keep raw material prices down.

90. Uneven development: consequences

One advantage is that the pictogram gives a very clear, highly visual indication of the inequality between regions. One disadvantage is that it is hard to know what quantity is represented when just a bit of a pictogram icon is used, as in the maternal deaths pictogram where the icon of a pregnant woman is sometimes presented as a half icon or a fraction of an icon.

91. Investment, industry and aid

A US$18.11 billion

92. Technology, trade, relief and loans

A premium is an addition to the market price; something that increases the price of the fair trade product by a set amount. The premium helps reduce the development gap because a set amount of this premium goes back to the producers. Most fair trade schemes require producers to use this money in specified ways, for example to increase wages of people working for them, to spend on community projects and to improve their businesses.

93. Tourism

(i) **Help local people to benefit more from tourism** – some international tourism has little benefit for local communities because tourists stay in self-contained, all-inclusive resorts: they rarely go outside their resorts and do not buy anything from local traders. If the resorts are owned by non-Kenyan companies – sometimes large TNCs – tourism will not reach its potential for reducing the development gap.
(ii) **Tourists encouraged to be more responsible** – tourists can cause damage to the attractions they are coming to see: for example, by dropping litter, being careless and causing fires or offending local people in the way they act or dress.
(iii) **Tourism companies encouraged to develop less damaging types of activities** – tourist companies can damage the environment (for example, with vehicles and soil erosion, or with unattractive, eyesore hotel developments) and make

places less attractive to future tourists, which could reduce tourist numbers and cause development gaps to widen.

94. LIC or NEE country: location and context

This answer will depend on the case study you are doing for this topic. For India, the change in the balance of sectors of the economy since 1990 looks like this:

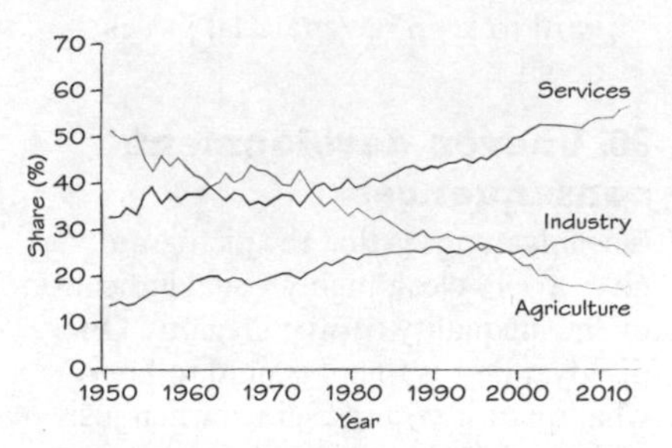

The real driver of economic development in India has been the growth of the service sector. Although India's agriculture has declined as a contributor to GDP, this is not because farming is less important than it used to be: in fact, India is now a net exporter of food while it used to rely on food imports. What has happened is that farming has modernised and become mechanised, so millions of people have left farming to work in industry and the service sector. This has caused economic development because all those people gave India a huge resource for making things, selling things and buying things. At the same time, Indian agriculture has become much more efficient.

95. LIC or NEE country: TNCs

This answer will depend on your case study country, but make sure that you include facts and figures specific to your chosen country.
This kind of question has marks for three assessment objectives (AOs): AO1, AO2 and AO3.

- For AO1, the best answers will show that you can use detailed knowledge of places and processes to support your answer.
- For AO2, the best answers will show that you understand the ways in which TNCs can contribute to industrial development, and the ways in which they can cause problems for a country's industrial development.
- For AO3, the best answers will show that you can apply your case study knowledge to this question and use it to decide 'to what extent'.

96. LIC or NEE country: trade, politics and aid

To find your case study country's top three trading partners, look online at The Observatory of Economic Complexity. This has excellent data visualisations for exports and imports for a wide range of countries.

97. LIC or NEE country: environmental impact

This answer will depend on the country you have studied for your case study. Here is an example answer for India.

Rapid economic development in India has led to serious water pollution problems. For example, the River Ganges has high levels of water pollution due to human waste and industrial waste. The river flows through more than 50 cities and 50 large towns. Many of these settlements use the river for disposing of sewage, much of it untreated, which is highly toxic to fish and plants.

Economic development is also increasing India's greenhouse gas emissions. India is currently the third largest emitter of CO_2 in the world, mainly due to its very large population (17 per cent of people in the world live in India). As India continues to develop, these emissions are likely to increase. At present, Indians emit 1.4 tonnes of CO_2 per person, compared to 17 tonnes per person in the USA.

98. UK: deindustrialisation, globalisation and policy

Your answer is likely to make the point that globalisation has enabled products, including coal, to be transported around the world at low cost. This means that countries that have large, easily accessible coalfields, efficient mining technology and a large workforce, who are prepared to work for lower wages than coalminers in the UK, are able to export coal to countries like the UK at a cheaper cost than coal produced in the UK. In the UK, deep coal mines are expensive to run and miners' wages are much higher than in LIC or NEE coal-producing countries. UK energy companies want to make profits and they will not buy expensive coal from the UK when cheaper coal is available from other countries.. That makes demand for UK coal too low to keep the deep mines open.

99. UK: post-industrial economy

This answer will depend on the example you have studied.
This kind of question is marked according to a mark scheme that will look at the ways you have applied your knowledge and understanding to answer the question.

Answers that are very general, perhaps simply describing one general way in which industries can reduce their environmental impacts, will not do well because they do not use specific information from a relevant example.

Answers that use good understanding of ways in which industries can reduce their environmental impacts will do better, but they will not reach the best marks unless they use specific information from a relevant example.

100. UK: rural change

Reasons are likely to include:

- to prevent the city 'sprawling' into the countryside in an uncontrolled way, making it difficult for city government to provide urban services: transport, schools, security, health
- to protect the villages in the green belt from losing their identity and character
- to protect the villages in the green belt from unsustainable growth with impacts on local services such as schools, health care and road use
- to protect farmland and food production in the UK
- to conserve natural habitats and biodiversity
- to provide green spaces for city residents to enjoy, and places where people can go for leisure activities, such as golf, biking, running and fishing
- to help with reducing carbon emissions.

101. UK: developments and differences

Your answer should make some or all of the following points.

- Without HS2, there are strong regional differences in house prices between the south-east and the rest of the country.
- The high house prices in the south-east are related to there being more job opportunities in this region, the highest wages and the highest demand for housing.
- House prices are lower in other parts of the UK because the demand for housing is lower. This is partly because job opportunities may not be as good here as in London and the south-east, because more businesses have located in the south-east. Wages may be lower in these regions.
- If HS2 were to reduce the journey time from northern cities to London (for example, 49 minutes from Birmingham to London, 1 hour 8 minutes from Manchester to London) then more people would decide to live in the northern cities and commute to London.
- They would make this decision because they would be able to afford housing in northern cities that would be too expensive for them in London or the south-east, and because the journey time to London would be an acceptable amount of time –

many people already make one hour journeys to London from all over the south-east.

- If more people decide to move to live in Birmingham or Manchester but work in London, house prices in these northern cities would increase because of increased demand and because these new residents would be earning higher London wages. This would start to reduce the regional differences in house prices seen in the map. Property prices in London might even begin to decrease as the pressure on available housing was reduced.

102. UK and the wider world

This is an evaluation question: you need to weigh up the political disadvantages against the economic advantages. Your evaluation needs to be detailed and balanced (so consider both advantages and disadvantages equally), then use your analysis to justify your final decision on whether you agree with the statement or not.

Political disadvantages could include the following.

Immigration (part of belonging to the EU's single market – an economic advantage) – a major reason for the UK voting against EU membership in 2016 was concern about the rapid increase in immigration from EU countries, especially from countries in Eastern Europe. This was political because it impacted on voting and also on the way the government decided the UK needed to be run.

Concerns about democracy – although individual countries elect representatives to the European Parliament, the leaders of the EU are not elected. There were concerns in the UK that important decisions were being made that the UK parliament had no control over.

Political concerns about the long-term direction of the EU – the EU was formed at the end of the Second World War with the aim of building so many links between the different countries of Europe that another war in Europe would be much less likely to occur. However, this meant that the long-term direction of the EU is 'ever-closer union'. Many people in the UK wanted to keep the UK independent, not part of what UK 'euro-sceptic' newspapers referred to as a 'European super-state'.

Political concerns about restrictions on what UK businesses were able to do – standard EU ways to run their business meant a lot more paperwork which businesses found problematic.

Economic advantages could include:

Access to the single European market – membership meant being able to trade with other EU countries without tariffs or other things that make trade more expensive.

The EU is the world's largest trading bloc – this means that the EU is able to negotiate trade deals with the rest of the world (including other trading blocs) that can be much more advantageous for members of the EU than the deals that those countries could make outside of the EU.

Immigration – although a political disadvantage, immigration from the EU was advantageous for businesses as it gave them a much bigger pool of people to recruit from, from the most highly skilled people to unskilled workers willing to work for low pay and in challenging working conditions.

Lower prices for consumers – the economic advantages of EU membership sometimes meant that consumers had cheaper prices for products and services (such as energy) while still benefitting from the protection on product quality and service quality that comes as part of being part of the EU.

103. Essential resources

Your answer is likely to relate to the level of development of the USA as the world's biggest economy, compared to the lower levels of development in many African countries. People in America have energy-intensive lifestyles (they drive cars, have air conditioning in summer and heating in winter, use many electrically powered devices. People in LIC countries in Africa do not have this type of lifestyle especially in rural areas where people walk or cycle or use buses rather than own their own cars, cook using wood for fuel and use low amounts of electricity. American farming and industry are highly mechanised, depending on large amounts of oil to generate power. Farming in LIC African countries uses a lot more human power rather than relying on machinery, industry is at a lower level than in HICs and demand for power is lower.

104. UK food resources

Your answer is likely to refer to the concerns that some UK consumers have about the agricultural chemicals used to increase yields, prevent disease and pest attack; organic food is grown without these chemicals. Ethical foods also includes foods that are produced in ways that minimise the suffering of animals; people who have concerns for animal welfare might choose these. Ethical foods also include foods that are produced in ways that reduce environmental damage (such as rainforest deforestation) or benefit local farming communities in LICs or NEEs. People who have concerns about environmental issues might choose these products.

105. UK water resources

Most electricity generation is powered by coal, gas or nuclear energy and all of these processes need water to provide cooling during the generation – water absorbs the excess heat and is then pumped away. Around half of all the water abstracted (taken) from UK rivers and the sea is used for cooling. So increasing demand for energy is likely to be linked to an increasing demand for water.

106. UK energy resources

Countries like the UK depend on affordable energy prices for UK consumers – otherwise there are negative economic impacts on jobs and households that are politically unpopular. However, importing energy from foreign countries has risks if those countries are politically unstable. Political tensions or problems could cause changes in supply or sudden price rises. Fracking in the UK would increase UK energy security as it would produce affordable energy from the UK's own resources, making the UK more self-sufficient in energy.

107. Demand for food

Your reason could include one of the following:

- As countries get wealthier, they can afford to import more food, which increases the range of food available and may also mean lower prices for consumers.
- When people have more money, they have more money to spend on food, for example, meat – something that poorer people cannot afford to eat very often. Processed food, which is often high in calories, also becomes more popular as people have less time to spend preparing food and turn to ready meals or pre-prepared meals.
- Wealthier countries may be able to subsidise agricultural production so that farmers can get a better price for their crops (increasing supply) without consumers having to pay more for their food.
- Economic development in agriculture means more mechanisation (which makes farming larger-scale and more efficient), higher inputs of fertilisers and pesticides (which improve yields) and increased farming expertise and use of scientific research, all of which means much more food produced more efficiently and therefore potentially more cheaply.
- Increased personal incomes often means that people eat a wider range of products, which boosts demand for agricultural production within the country, increasing economic development.

108. Food insecurity

Conflict has a negative impact on food supply. If people are unable to farm their fields for fear of attack, or because there is a danger of mines or unexploded bombs, or because farmers have been called up into armies or pressurised to fight, then the supply of food decreases. People often also attempt to flee conflict areas, meaning that farms are deserted, while large populations of refugees grow up in safer areas where the demand for food usually has to be met by international aid.

Food prices increase as people compete to buy scarce food, which can mean people becoming undernourished, especially in cities where food supplies are not able to get in from outside the city or in rural areas where soldiers loot or requisition farmers' crops and livestock. Governments that are engaged in conflicts may not have money to spare for importing crops and may be unable to send food aid to areas at risk of famine or food shortages if they are occupied by enemy forces.

109. Increasing food supply

This answer will depend on the example you have studied.

Advantages often include the following:

- Less water is used due to efficient irrigation methods, including hydroponics, computer-regulated drip irrigation.
- Training may be provided for local farmers to help them act as suppliers to the large-scale agricultural development; this may improve yields for the whole region.
- Large-scale irrigation schemes can also benefit many local farmers, improving yields.
- Large agricultural companies may build houses for workers, possibly also schools and health care facilities.
- Jobs are created in the local area, for seasonal workers.
- Cheaper prices for crops grown due to efficient methods and economies of scale.
- The diets of local people can improve when they have access to a wider range of foods.
- Processing plants may be set up in the local area, creating more jobs.
- Food exports may increase as a result of the large-scale agricultural development, helping national GNI.
- Increased food production can mean improved food security for the region and the country.
- Large-scale agricultural developments often invest in their own power generation, which often involves sustainable methods, such as solar power or mini HEP projects.

Disadvantages often include the following:

- There can be high costs and high energy demands involved in developments that use advanced technology.
- Wages for agricultural workers can be very low, and an influx of seasonal migrant workers can cause tensions with local people.
- Local farmers who act as suppliers may end up in debt as they borrow to buy GM seeds.
- Large-scale agricultural developments use up huge areas of land that was previously farmland for local people and often included villages where people used to live – alternative accommodation and farmland may not be as good as what people had before.
- The reduction in local land availability may mean rents go up for small farmers.
- Impacts on the local environment, for example, large areas covered by glass greenhouses or plastic greenhouses.
- Irrigation can cause increased salinisation and waterlogging of the soil, reducing yields over time.
- Increased use of pesticides can have impacts on habitats outside the agricultural development, reducing local biodiversity.
- Runoff from the large agricultural development may contain high levels of fertiliser or other chemicals, such as herbicides, which can damage local habitats.
- Impacts on local groundwater supplies or river levels due to increased abstraction of water for irrigation. Dams for HEP schemes can reduce the amount of water reaching farmers downstream.
- Impacts on microclimates: large areas of greenhouses reflect heat back into the atmosphere.

110. Sustainable food supplies

Your answer will depend on the example you have studied. Ways of increasing sustainable supplies of food commonly include the following:

- Benefitting from government agricultural research projects and non-governmental organisations' research and expertise.
- Increasing water supplies through rainwater harvesting/dams.
- Improving water quality by filtering it through sand or reed beds.
- Education for local farmers on ways to improve yields in sustainable ways (for example, through composting, mixed agriculture where the manure from livestock is used to maintain soil nutrient levels, techniques like rice-fish farming, where fish are farmed in the water-filled rice paddies, with the fish eating pests and providing fertiliser through their excrement, and also providing a source of protein for the farmers).
- Reforesting slopes and growing trees among crops (agroforestry) to act as wind breaks, to stabilise the soil and increase biodiversity; also growing a mixture of crops (such as maize and beans) to provide nitrogen-fixing for the soil.
- Providing land plots for homeless and landless people and families, slowing rural–urban migration.
- Developing urban farming as a way of helping the urban poor to improve their diets, find work and get affordable nutritious food.
- Involving whole communities in farming improvement schemes and providing local people with jobs as trainers, maintenance teams, and so on.

111. Demand for water

Your reason could include one of the following:

A rising population – measure of water scarcity is water stress, measured in terms of water availability per person. Since water is not a resource with any alternatives (unlike oil, for example), the more people there are who depend on the same supply of water, the less water there is per person.

Economic development – economic development, especially industrialisation, means a huge increase in water consumption because of the high demand for water from all economic activities. One example is that, as a country develops economically, people tend to move from a basic cereal and vegetable diet to one containing more meat, and livestock farming uses much more water than farming crops.

112. Water insecurity

Food production – although drought-tolerant varieties of crops are being developed, and farmers can switch from crops that require a lot of water to crops that require less (such as millet), water insecurity makes farming much more difficult and unpredictable. Water insecurity is linked to rising food prices, as there is less food in water-insecure places, reducing supply, which pushes prices up, and rising food prices put huge pressures on poor people.

Conflict – as well as increasing stress on people and contributing to rising food prices, which are closely linked to conflict, competition between regions and between countries has the potential to lead to large-scale conflicts. A major potential source of conflict are dam projects: where a country upstream decides to improve water security by constructing a dam, water insecurity often increases for countries downstream, causing international conflict, as with the Nile.

Water pollution – the impacts of water pollution are increased by water insecurity as people may not have alternative supplies of unpolluted water to use instead of the polluted water. Water security includes safeguarding populations from waterborne pollution and water-related disasters.

113. Increasing water supply

This answer will depend on the example you have studied. Many students study the South-to-North Water Transfer Project (SNWTP) in China, which is the example used here.

Advantages include the following:

- The transfer of huge volumes of water from southern China, which has 77 per cent of China's water resources, to the drier north. The huge growth of industry in the north and its rapid urbanisation have meant growing water insecurity for agriculture. The SNWTP aims to transfer 12 trillion gallons of water 1000 miles from south to north.
- When completed, the plan is for the SNWTP to ensure water security for 500 million people, allowing cities like Beijing to stop the over-abstraction of groundwater resources and increasing the economic development opportunities of China's dry north.
- Water will no longer have to be diverted from farmland to meet the industrial and domestic needs of China's northern cities and factories. This will mean greater water security for the 134 million people farming in water deficit regions in China's north.

Disadvantages include the following:

- **Cost** – this enormous engineering scheme has an estimated price tag equivalent to US$62 billion. Currently, China's economic growth is slowing and the budget for the SNWTP may become challenging to justify.
- **Social impact** – 300 000 people had to move in order for the construction of a reservoir along the central route of the SNWTP.
- **Environmental impact** – the parts of the scheme already completed have suffered from water pollution problems, both agricultural and industrial pollution. By the time the water gets to the northern cities, it is often too polluted to use.
- **Environmental impact** – because the water is mainly transferred in canals and open channels, evaporation losses are huge. The water transfer is not efficient, considering the huge cost of moving the water in the first place.
- **Environmental impacts** – climate change is a threat to the scheme because it may make droughts more common in the south – there is already some evidence of this happening. Rainfall has actually increased in the north, undermining the rationale for the transfer scheme.

114. Sustainable water supplies

This answer will depend on the example you have studied. Ways of increasing sustainable supplies of water commonly include the following:

- Increasing water supply by collecting more rainwater: for example, using appropriate technology.
- Increasing water supply by storing more rainwater: for example, underground storage tanks.
- Increasing water supply by installing low, concrete dams (sand dams) in river beds (during the dry season) or stream beds on hillsides to create mini reservoirs. The low river dams trap sand and sediment behind them, water sinks into these deposits and is stored there for use in the dry season. The sand also filters the water, making it safer to use.
- Small-scale water transfer schemes to pipe water from wetter upland regions or upland springs down to drier valleys where people farm. Using gravity for this water transfer rather than fossil-fuel-powered pumps is more sustainable.
- Increasing water supply through digging irrigation ditches linked to bunds – mini dams that are put into rivers in the wet season to channel water down the irrigation ditches to one group of fields, then the bund is moved downstream to the next channel and the next set of fields.
- Increasing water conservation: for example, introducing appropriate technology such as drip irrigation systems, installing covers on wells and storage tanks to reduce evaporation.
- Increasing understanding of rainfall measurement – when farmers measure rainfall it enables them to recognise the best times to plant crops, and which crops to plant.
- Putting water management schemes into the hands of local communities is sustainable. The communities will then manage them to meet their needs and will be able to make necessary repairs and maintenance.

115. Demand for energy

1 0.2 billion metric tons of oil equivalent

2 Explanations might include:

- China's rapid industrialisation, which has also increased China's demand for oil to power its factories and to be used in the manufacture of its products
- China's economic development, which has meant millions more Chinese each year can afford to buy a car or can afford to travel further by car, increasing consumption of oil
- it highlights China's lack of its own oil resources.

116. Energy insecurity

1 **Exploration of difficult areas** – areas like the Arctic are challenging for energy exploitation but energy insecurity means countries will be keen to exploit them once technological advances or price changes makes it economically viable to do so.

Exploration of environmentally sensitive areas (another aspect of exploration of difficult areas) – these are often environmentally sensitive areas, which could easily be damaged by energy exploitation. Although energy insecurity makes exploring environmentally sensitive areas attractive for energy companies, they often have to deal with environmental pressure groups. For example, in 2015 the energy company Shell pulled out of oil and gas exploration in the environmentally sensitive cold environment of Alaska, due partly to lower-than-expected oil and gas finds, and partly to very strong environmental protests.

Economic costs – growing energy insecurity will often compel a country to spend money on reducing its insecurity. Nuclear energy is a good example: the investment required is enormous, but governments make the argument for this economic cost by warning that such investment is needed 'to keep the lights on' in the country.

Environmental costs – although a country might wish to protect its environment, energy insecurity has such serious implications for its economy and political situation that environmental concerns have to be sacrificed if there is energy to be exploited. An example would be tar sand exploitation in Canada. Canada is a country that takes its environmental responsibilities seriously. However, it has invested significantly in tar sand exploitation, although there are significant risks of major environmental damage from this form of energy extraction.

Potential for conflict – energy insecurity is a major cause of conflict. Countries like Russia which have large energy surpluses use their control over supply to put pressure on other countries. Countries like the USA have intervened militarily in oil-producing countries to make sure that energy supplies remain reliable. International organisations also impose sanctions on countries that are behaving in unacceptable ways to stop them importing energy if they are energy-deficit countries, or to stop them exporting energy if they are energy-surplus countries.

2 Your answer is likely to consider two main factors:

- how increases in oil and gas prices make extracting oil and gas from challenging environments more attractive
- how countries with energy supplies in challenging places may decide to exploit them in order to make at least some of their energy supply less dependent on other countries.

It would be a good idea to start by explaining that challenging environments make extracting oil and gas much more expensive (for example, dealing with deep water, very cold temperatures, remote locations, difficult climate conditions such as severe winter storms). However, when political or economic issues cause big increases in the price of oil and gas, the costs of working in challenging environments no longer mean that production there would be unprofitable.

For the second point, the explanation would be along the lines of: the cost of production in challenging environments may be too high to make production reliably profitable, and there may be considerable local or national opposition to extracting this energy (perhaps for environmental reasons). Even so, governments may decide to go ahead because, strategically, the country needs to reduce its reliance on other countries, especially if those countries are not traditional allies or have problems with political stability.

117. Increasing energy supply

This answer will depend on the example you have studied.

Advantages may include the following:

- If a country has sizeable fossil fuel deposits, this increases the country's energy security: for example, the UK's North Sea oil and gas reduced the UK's dependence on foreign energy suppliers.
- Fossil fuels can mean that a country is able to export energy. For some fossil-fuel-rich countries this is what the majority of their economic development depends on.
- Fossil fuels are very effective sources of power and allow both agricultural and industrial development.
- Fossil fuel extraction creates a lot of jobs: in 2008, 450 000 UK jobs were directly in North Sea oil and gas or in industries linked to it.
- If a country has relied on fossil fuels for a number of years, it already has the infrastructure to store these fuels, process them, transport them, use them to generate power and then transfer that power from power stations to homes and businesses across the country.
- Although some fossil fuels are heavy polluters and carbon emitters, some are cleaner than others: for example, natural gas produces 45 per cent less carbon dioxide than coal.

Disadvantages may include the following:

- Burning fossil fuels releases carbon dioxide into the atmosphere: a greenhouse gas.
- Fossil fuels are finite – they eventually run out. For example, the UK's reserves in the North Sea reached peak production in 1999; production is now gradually declining.
- Fossil fuel prices drive the world economy and a fall in prices can hit the economic development of oil-exporting countries very hard.
- Extracting fossil fuels is dangerous work and the risk of accidents needs to be carefully monitored. Accidents can kill workers and can also cause very significant environmental disasters such as oil spills. In 2010, the Deepwater Horizon oil disaster in the Gulf of Mexico killed 11 people and released 4.9 million barrels of oil into the marine ecosystem of the Gulf of Mexico. The oil company, BP, had to pay US$18 billion in fines because of the environmental damage caused.
- Most countries have signed up to treaties committing them to lowering carbon emissions. To do this, countries have to reduce their reliance on fossil fuels and increase their use of alternative, sustainable energy sources.

118. Sustainable energy supplies

This answer will depend on the example you have studied.

LIC or NEE local renewable energy schemes often involve micro-hydro. These tend to be set up by a non-governmental organisation (NGO) working closely with local communities. Micro-hydro schemes are small scale and powered by gravity. They are suitable for communities in mountain areas where the steep relief and fast-flowing streams can be used to power a small turbine and generate renewable energy. Many micro-hydro schemes do not require dams and reservoirs; they divert a stream into a settling tank that removes debris (which might otherwise damage the turbine). Then the water falls down a pipeline to power the turbine. This type of micro-hydro is called a run-of-the-river system.

Often the local community has to pay towards the cost of the scheme, with government and NGO assistance, but run-of-the-river systems are low-cost, especially when compared to larger-scale hydro projects. If local people are trained in operating and maintaining the micro-hydro plant, there are usually only low onwards costs for running and repairing the plant and the whole community is able to benefit from the electricity generated by the scheme. Homes have heat in the winter and light at night-time, people can have TVs and refrigerators. New industries can be set up, helping the economic development of the community.

Other types of local renewable energy schemes include solar power mini grids and local biomass schemes. If your answer involves these, make sure you have described how the scheme generates renewable power to be a sustainable energy source and what makes it local (for example, the scale of the project, the people that it benefits. Also make sure you have clearly identified which LIC or NEE it is located in.

122. Enquiry questions

This answer will be specific to your fieldwork enquiry: you would have different answers for your human enquiry and for your physical enquiry. One feature that makes fieldwork locations appropriate is that they are comparable. If you want to sample how something changes, you need to try to keep all other factors as similar as possible. Other factors that are important include accessibility, safety and whether useful secondary data are easily available.

123. Selecting, measuring and recording data

Quantitative: Measuring water quality by sampling
Qualitative: Collecting views on river flooding by questionnaire

124. Processing and presenting data

This will depend on your geographical enquiry and the method or methods that you used. Here is an example answer for a student comparing traffic counts into and out of a coastal village in an enquiry on the impacts of tourism.

I used a comparative line graph to present the findings of the traffic counts. There were four lines on the graph for the four different counts, with the times of the counts (from 9.30am to 3.30pm) along the horizontal axis. I chose this presentation method because the data was continuous (times through the day) and because I wanted to be able to easily identify any variations between the four counts.

125. Analysing data and reaching conclusions

$3.0 + 2.8 + 3.9 + 2.0 + 1.0 + 2.6 = 15.3 \div 6 = 2.6$ (to 1 decimal place)
Remember: when calculating the mean, you need to add all the numbers together and then divide by the number of variables.

126. Evaluating geographical enquiries

Limitations of quantitative data include: it can be abstract and hard to relate to particular local situations; it can oversimplify what is actually a complex situation; it is only as good as the categories that the researcher decides to use in collecting the data.

Limitations of qualitative data include: it is difficult to categorise (different people have different views), which makes it hard to analyse; it is difficult to generalise (just because 10 per cent of people in an area have a particular view does not mean 10 per cent of people will have that same view in the country as a whole); it can be challenging to make it representative (complicated sampling methods are required).

127. 6-mark questions

To find supporting evidence for this question, look through the Sample Examination Questions (SAMs) or past question papers for this course on the AQA website, and write down the different types of 6-mark question that you see there.

128. 9-mark questions

To answer a 'to what extent' question you must weigh up the argument for and against the proposal. What are the reasons for thinking 9-mark questions are the hardest? Why might other types of AQA GCSE geography questions be harder, or just as hard?

For example, look at the following question.
To what extent does a cold environment you have studied provide both opportunities and challenges for development? **(9 marks)**

Here you need to:

- use your case study knowledge accurately and relevantly to evaluate in detail several opportunities of development for cold environments
- then do the same for several challenges for development
- then consider with justification from your case study knowledge, whether you agree that opportunities and challenges for development are quite evenly balanced, or whether (as is probably more likely) either the challenges or the opportunities are more important than the other.

129. Paper 3

A = 2; **B** = 1; **C** = 3

130. Atlas and map skills

The distribution shows that tropical rainforests are found only in a narrow zone centred on the Equator. This suggests that they need warm temperatures and low seasonal variation to develop. This is also an area of the global atmospheric circulation system dominated by the Intertropical Convergence Zone, bringing high rates of precipitation. In order to explain why there is not a continuous belt of tropical rainforest you could talk about local factors: For example, high mountains in South America to explain why tropical rainforests do not extend completely from east to west over the continent. Human activity could be another reason – deforestation – to explain why distribution within this tropical zone is uneven. More than two reasons have been provided here but remember in the exam to only provide the number of reasons requested.

131. Types of map and scale

A A lowland river on a wide flood plain

132. Grid references and distances

The 1:25 000 scale shows more detail. It may be confusing because 1:50 000 sounds bigger. But what the scale means is that an object at 1:25 000 is 25 000 times smaller than it is in real life, while the same object at 1:50 000 is 50 000 times smaller.

133. Cross sections and relief

1

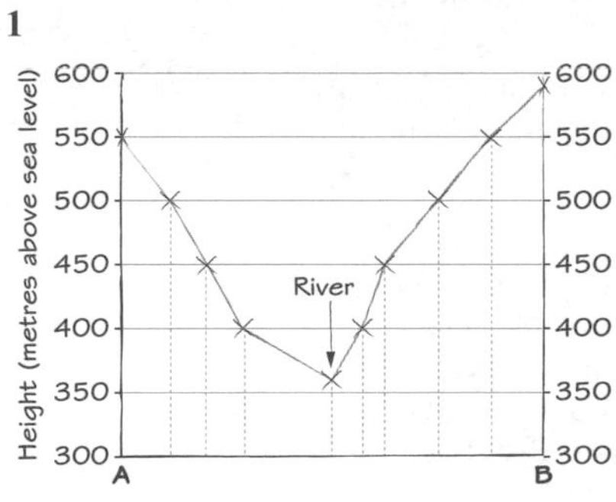

2 While contours link points on a map of the same height, spot heights give an exact height above sea level. They are shown by a black dot on the map with a figure next to it which is the height above sea level of that point on the map. The figure is in metres.

134. Physical and human patterns

Settlement Y is located at a crossroads, so access is easy, which encourages development of a nucleated settlement. It is also located close to another large area of development, settlement Z, so settlement Y could be a commuter area. While settlement Z has limited space to expand because of its location in a meander loop (and settlement X is restricted by mountains) settlement Y does not have restrictions and can expand more easily.

135. Human activity and OS maps

Your answers could include two from:

- at Cider Farm (567246) there is a map symbol for caravanning and camping
- in Ross-on-Wye there is a tourist information centre (599242), next to a viewpoint
- hotels are indicated at the Wilton roundabout on the outskirts of Ross-on-Wye (587243)
- the remains of a castle are indicated at 593244.
- footpaths are shown along the river, which could be popular with tourists.

136. Sketch maps and annotations

An advantage of annotated diagrams are that they are effective for showing change over time – for example, in beach profiles or river channel characteristics – which is often difficult to explain in words alone.

137. Using and interpreting images

Your two descriptions could include points like: identifying developing tropical storms in a source area, monitoring the track of a tropical storm over time, monitoring the speed of tropical storm movement, estimating likely landfall(s), gauging the damage done by tropical storms after landfall.

138. Graphical skills 1

(a) line graph; (b) scattergraph; (c) pie chart

139. Graphical skills 2

River flow or river discharge

140. Graphical skills 3

An isoline map – the contour lines join up heights of equal value

141. Numbers and statistics 1

162.5%

Population increase is worked out by:

- finding the increase:
  new number − original number

so 21 million − 8 million = 13 million

- dividing the increase by the original number and × 100

so 13 million ÷ 8 million = 1.625 × 100 = 162.5%

142. Numbers and statistics 2

The interquartile range is the difference between the upper quartile and the lower quartile. The lower quartile is worked out on the page as being 5 and the upper quartile as 11.
11 − 5 = 6

Published by Pearson Education Limited, 80 Strand, London, WC2R 0RL.

www.pearsonschoolsandfecolleges.co.uk

Typeset and illustrated by Kamae Design, Oxford
Produced by Out of House Publishing
Picture research by Ruth Smith

Cover illustration by Miriam Sturdee

First published 2017

20 19 18 17
10 9 8 7 6 5 4 3 2

British Library Cataloguing in Publication Data
A catalogue record for this book is available from the British Library

ISBN 978 1 292 13132 0

Printed in Slovakia by Neografia

Acknowledgements
Content is included from Michael Chiles, David Flint, Anne-Marie Grant and Kirsty Taylor.

The author and publisher would like to thank the following individuals and organisations for permission to reproduce copyright material:

Photographs

(Key: b-bottom; c-centre; l-left; r-right; t-top)

Alamy Stock Photo: AfriPics.com 36cr, Andrew Woodley 92, Ashley Cooper 67, Blend Images 65l, blickwinkel 40, Chris Gibson 61, christopher jones 104bl, Convery flowers 85, Darrin Jenkins 99, David Cole 23c, David Gowans 136, Dinodia Photos 75, FLPA 23t, 123bc, frans lemmens 62, Graham Taylor 23cr, imageBROKER 137bl, imagegallery2 12, incamerastock 23cl, Izel Photography 100, Jack Sullivan 50, Jo Katanigra 66, John Morrison 68br, John Smith 84, John Sylvester 41cr, johnrochaphoto 41bl, nik wheeler 33, Oyvind Martinsen 42, PA Images 98, Paul White - North West England 58, Philip Scalia 13, PhotoStock-Israel 114, Rehman Asad 74, Richard Burdon 63, Russell Watkins 123br, Simon Whaley Landscapes 68cr, Stacy Walsh Rosenstock 110, Stephen Dorey ABIPP 70, Terry Whittaker 52tr, Tim Graham 69, Tom Francis 46, Tom Gilks 36cl, TravelStockCollection – Homer Sykes 34; **Digital Vision:** 27; **Fotolia.com:** Cloudia Spinner 137tr, ian woolcock 49tl, miket 137tl, numage 49tr; **Getty Images:** Arthur Morris 24, Geography Photos 55, George Steinmetz 22, 35, Joe Raedle 14, Justin Sullivan 6tr, Keren Su 95, Matt Cardy 15cr, Richard Baker 83, Sam Diephuis 21, Sven Zacek 38, Yva Momatiuk & John Eastcott / Minden Pictures 43; **RainSaucers Inc.:** 37; **Science Photo Library Ltd:** Digital Globe, Eurimage 6br, Getmapping PLC 52bl, NASA 137cl, NOAA 131, Photostock-Israel 113; **Shutterstock.com:** 724258 1, Alf Ribeiro 117, Bakhtiar Zein 119, BNMK0819 121, Christian Bertrand 94, Jenny Solomon 79, John Gomez 106, Kevin Eaves 44cl, monotoomono 76, Nick Hawkes 44tr, Nickolay Vinokurov 137cr, northallertonman 15tl, Sozaijiten 65cr, Svetlana Serebryakova 104tr; **UNAVCO:** 7

All other images © Pearson Education

Maps

Map on page 17 from Provisional rainfall percent of average map for Winter 2015/16 (December, January and February), Met Office, contains public sector information licensed under the Open Government Licence v3.0; Ordnance Survey maps on pages 44, 50, 57, 58, 59, 65, 68, 82, 131, 132, 134, 135 © Crown copyright 2017, OS 100030901 and supplied by courtesy of Maps International, created by Lovell Johns Limited, Maps International is a trading name of Lovell Johns Ltd.

Figures

Data for the graph on page 11 is from Climate Change Indicators: Tropical Cyclone Activity, Figure 1. Number of Hurricanes in the North Atlantic, 1878–2015, US Environmental Protection Agency; figure on page 85 is adapted from the graph on page 3 of Influencing the growth of cycling in London, TfL Integrated Cycling Research Group; figure on pages 90 and 120 is from WHO, World Health Statistics 2016, reprinted from World Health Statistics, page 45, Copyright 2016; data for the graph on page 90 is from: House of Commons Library briefing paper Migration Statistics, Number SN06077, 26 May 2016 p. 14 (chart 8); figure on page 96 © 2014 Euromonitor, by permission of the author Virgilijus Narusevicius and Euromonitor International; graph on page 106(t) from Renewables overtake coal; Fuels used for electricity generation graph, Department of Energy & Climate Change, contains public sector information licensed under the Open Government Licence v3.0, © Crown copyright.

Note from the publisher
Pearson has robust editorial processes, including answer and fact checks, to ensure the accuracy of the content in this publication, and every effort is made to ensure this publication is free of errors. We are, however, only human, and occasionally errors do occur. Pearson is not liable for any misunderstandings that arise as a result of errors in this publication, but it is our priority to ensure that the content is accurate. If you spot an error, please do contact us at resourcescorrections@pearson.com so we can make sure it is corrected.